Praise for *Spiritual Science*

"As I read this book, I kept sighing with relief. At long last, a thoughtful and accessible treatment of the false divide between science and spirituality. By exploring a series of puzzles, Taylor shows how the pieces of our world fit together, if we are willing to take a breath and look at it anew."

Dr Julia Mossbridge, author of
Transcendent Mind and *The Premonition Code*

"With elegance and lucidity, Steve Taylor explains why spiritual science is the only hope for humanity. A science based on the superstition of matter as fundamental reality could lead to our extinction but a science grounded in the understanding of consciousness as a fundamental reality – as described by this book – could be our saving grace."

Deepak Chopra,
MD Chopra Foundation

"In this important book, Steve Taylor convincingly argues that the materialist paradigm has run its course and that the evidence from anomalous experiences must be acknowledged. Taylor shows how a panspiritist approach not only eloquently explains anomalous phenomena but can lead to exciting possibilities for the evolution of humankind and the planet. These issues affect each one of us; it is time we all sat up and took note."

Dr Penny Sartori, author of
The Wisdom of Near-Death Experiences

"Materialism is dead. It just doesn't know it. *Spiritual Science* shows the mechanistic worldview is passé and that the science that once seemed to support it has well and truly moved on. Steve Taylor's book is a very readable and inspiring guide to where we are heading as a culture."

Gary Lachman, author of
Lost Knowledge of the Imagination

Steve Taylor PhD is a senior lecturer in psychology at Leeds Beckett University, and the author of several bestselling books on psychology and spirituality. For the last seven years he has been included in Watkins' *Mind Body Spirit* magazine's list of the "100 most spiritually influential living people". His books include *Waking From Sleep, The Fall, Out of the Darkness, Back to Sanity* and his latest book *The Leap.* His books have been published in 19 languages, while his articles and essays have been published in over 40 academic journals, magazines and newspapers, including *Philosophy Now, Tikkun, Journal of Humanistic Psychology* and *Journal of Consciousness Studies.*

SPIRITUAL
SCIENCE

Why science needs spirituality
to make sense of the world

STEVE TAYLOR

WATKINS
Sharing Wisdom Since 1893

This edition first published in the UK and USA 2018 by
Watkins, an imprint of Watkins Media Limited
Unit 11, Shepperton House
89-93 Shepperton Road
London
N1 3DF
enquiries@watkinspublishing.com

1 3 5 7 9 10 8 6 4 2

Typeset by Integra Software Services Pvt. Ltd, Pondicherry

Printed and bound in the United Kingdom

A CIP record for this book is available from the British Library

ISBN: 978-1-78678-158-1

www.watkinspublishing.com

CONTENTS

CONTENTS

INTRODUCTION

As an academic – a researcher and senior lecturer at a university in the UK – people are often surprised by my unorthodox views on the nature of life, and of the world. For example, when I mention to colleagues that I'm open-minded about the possibility of some form of life after death, or that I believe in the possibility of paranormal phenomena such as telepathy or precognition, they look at me as if I've told them I'm going to give up academia and become a professional footballer. It's taken for granted that if you're an academic or an intellectual, you don't entertain such unusual views.

The majority of my colleagues and peers – and most academics and intellectuals in general – seem to have an orthodox materialist view of the world. They believe that human consciousness is produced by the brain, and that when the brain ceases to function consciousness will end. They believe that phenomena such as telepathy and precognition belong to a pre-rational superstitious worldview, which has long been superseded by modern science. They believe that the evolution of life – and most human behaviour – can be completely explained in terms of principles such as natural selection and competition for resources. To doubt these beliefs is to be seen as weak-minded or intellectually gullible.

People are even more confused when I tell them that I'm not religious. "How can you believe in life after death without being religious?" they wonder. "How can you be doubtful about Darwinism without being religious?"

This book is my attempt to justify my views to anyone who believes that to be rational means that by definition you also ascribe to a materialist view of the world. It's my attempt to show that one can be an intellectual and a rationalist without automatically denying the existence of seemingly "irrational" phenomena. In fact, I will show that it is actually much more rational to be open to the existence of such phenomena.

Beyond religion and materialism

Another aim of this book is to show that, although we might not be aware of it, our culture is in thrall to a particular paradigm or belief system that in its own way is just as dogmatic and irrational as a religious paradigm. This is the belief system of materialism, which holds that matter is the primary reality of the universe, and that anything that appears to be non-physical – such as the mind, our thoughts, consciousness or even life itself – is actually physical in origin, or can be explained in physical terms.

I hope to show that we don't just have to choose between an orthodox materialist view of the world and an orthodox religious view. Often it is assumed that these are the only two options. Either you believe in heaven and hell, or you believe that there is no life after death. Either you believe in a God who overlooks and controls the events of the world, or you believe that nothing exists apart from chemical particles and the phenomena – including living beings – that have accidentally formed out of them. Either God created all life forms, or they evolved accidentally through random mutations and natural selection.

But this is a false dichotomy. There is an alternative to the religious and materialist views of reality, which is arguably a more rational option than both. Broadly, this alternative can be termed "post-materialism".[1] Post-materialism holds that matter is not the primary reality of the universe, and that phenomena such as consciousness or life cannot be wholly explained in biological or neurological terms. Post-materialism holds that there is something more fundamental than matter, which might be variously termed mind, consciousness or spirit.

There are a number of varieties of "post-materialism". One of the most popular is called panpsychism, which is the idea that all material things (down to the level of atoms) have a degree of sentience, or consciousness, even if it is infinitesimally small, or just a kind of "proto-consciousness". However, for reasons that I will describe in detail in Chapter 2, I favour what I call a "panspiritist" approach. Or you could simply call it a "spiritual" approach.

The basic idea of my spiritual approach is very simple: the essence of reality (which is also the essence of our being) is a quality that might be called spirit, or consciousness. This quality is fundamental and universal; it is everywhere and in all things. It is not unlike gravity or mass, in that it was embedded into the universe right from the beginning of time, and is still present in everything. It may even have existed *before* the universe, and the universe can be seen as an emanation or manifestation of it.

Although this is a simple idea, it has a lot of important corollaries and consequences. Since all things share this common spiritual essence, there are no separate or distinct entities. As living beings, we are not separate from each other, or from the world we live in, since we share the same nature as each other and the world. It also means that the universe is not an inanimate, empty place, but a living organism. The whole cosmos is imbued with spirit-force, from the tiniest particles of matter to the vast, seemingly empty tracts of darkness between planets and solar systems.

Spirituality isn't often thought of in an "explanatory" context. Most people believe that it is the role of science to explain how the world works. But in this book we'll see that this simple notion – that there is a fundamental spirit or consciousness that is ever-present and in everything – has great explanatory power. We will see that there are many issues that don't make sense from a materialist perspective, but which can be easily explained from a spiritual point of view.

This is perhaps the biggest problem with materialism: that there are so many phenomena that it can't account for. As a result, it is woefully inadequate as a model of reality. At this point, it is reasonable to say that, as an attempt to explain human life and the world, it has failed. As I will point out throughout this book, only a worldview based on the idea that there is something more fundamental than matter can help us to make sense of the world.

The difference between science and scientism

One thing I would like to make clear at the beginning of this book is that I am not criticizing science itself. This is one of the common reactions I've had to the articles I've published on similar themes to this book. "How can you criticize science when it has done so much for us?" is a typical comment. "How can you tell me it isn't true when it's based on millions of laboratory experiments, and its basic principles are used in every aspect of modern life?" is another. A further typical query is: "Why do you equate science to a religion? Scientists don't care about beliefs – they just keep their minds open until the evidence appears. And if they have to revise their opinions, they do."

I have no wish to criticize the many scientists – such as marine biologists, climatologists, astronomers or chemical engineers – who work diligently and valuably without being particularly concerned with philosophical or metaphysical issues. Science is a method and process of observing and investigating natural phenomena, and reaching conclusions about them. It's a process of uncovering basic principles of the natural world, and of the universe, or of the biology of living beings. It's an open-ended process whose theories are – ideally – continually tested and updated. And I completely agree that science has given us many wonderful things. It's given us amazingly intricate knowledge of the world and of the human body. It's given us vaccinations against diseases that killed our ancestors, and the ability to heal a massive array of conditions and injuries that would also have been fatal in the past. It's given us air travel, space travel and a whole host of other incredible feats of engineering and technology.

All of this is wonderful. And it's partly because of such accomplishments that I love science. The other main reason I love science is that it opens us up to the wonders of nature and the universe. In particular, I love biology, physics and astronomy. The complexity of the human body – and particularly of the human brain, with its 100 billion neurons – amazes me. And I find it mind-boggling that we know the structure of the tiniest particles of matter, and at the same time have knowledge of the structure of the universe as a whole. The fact that scientific discoveries range from such a microcosmic level to such a macrocosmic level is incredible. I feel immense gratitude to the scientists throughout history who have made our present understanding of the universe and the world possible.

So why am I so critical of science? you might ask.

The answer is that I'm not critical of science or scientists. I am critical of the materialistic worldview – or paradigm – that has become so intertwined with science that many people can't tell them apart. (Another possible term for this is scientism, which emphasizes that it is a worldview that has been extrapolated out of some scientific findings.) Materialism (or scientism) contains many assumptions and beliefs which have no basis in fact, but which have authority simply because they are associated with science.

One of these assumptions is that consciousness is produced by the human brain. However, there is no evidence for this at all – despite decades of intensive investigation and theorizing, no scientist has even come close to suggesting how the brain might give rise to consciousness. It's simply assumed that the brain must give rise to consciousness because there appear to be some correlations between brain activity and consciousness (for example, when my brain is injured, my consciousness may be impaired or altered) and because there doesn't appear to be any other way in which consciousness could possibly arise. (In fact, as we will see in Chapter 3, there is a growing awareness of how problematic this assumption is, with more and more theorists turning towards alternative perspectives, such as panpsychism.)

Another assumption is that psychic phenomena such as telepathy or precognition cannot exist. Similarly, anomalous phenomena such as near-death experiences or spiritual experiences are seen as brain-generated hallucinations. Materialists sometimes say that if psi phenomena really did exist, they would break the laws of physics, or turn all the principles of science upside down. But this is untrue. As we will see later, phenomena such as telepathy and precognition *are* compatible with some of the laws of physics. In addition, there is a great deal of empirical and experimental evidence to suggest that they are real.

However, some materialists have a blanket refusal to consider the evidence for these phenomena, which is similar to the way many religious fundamentalists refuse to consider evidence that goes against their beliefs. This refusal isn't based on reason, but on the fact that these phenomena contravene their belief system. (This contradicts the naive assumption that science is always purely evidence-based, and theories and concepts are always re-evaluated in the light of new findings. This is how science *should ideally be*, but unfortunately any findings or theories that contravene the tenets of scientism are often dismissed out of hand without being given a fair hearing.)

Thankfully, there are some scientists who actively oppose materialism – scientists who have the courage to risk the hostility and ridicule of their orthodox peers and investigate potentially "heretical" possibilities, such as that there may be more to evolution that just random mutations and natural selections, that so-called paranormal phenomena may in fact be "normal", or that consciousness isn't wholly dependent on the brain. Heretical scientists aren't burned at the stake, of course, as religious heretics sometimes were, but they are often excommunicated – that is, ostracized and excluded from academia, and subjected to ridicule.

So in this book, I certainly don't intend to throw science overboard, and return to ignorance and superstition – far from it. I would simply like to free science from the straitjacket of the belief system of materialism, and as a result introduce a wider and more holistic form of science, one that is not limited and distorted by beliefs and assumptions – a *spiritual science*.

The structure of this book

This book begins by looking at the main principles of both materialism and panspiritism. Then I will take you on a detailed tour of a number of areas of scientific enquiry, during which I will highlight many problematic issues – or puzzles – which materialism struggles to solve. We will see that there are two ways in which the conventional materialist model of reality is deficient. One is that it cannot adequately explain major scientific and philosophical issues, such as consciousness, the relationship between the mind and brain (and the mind and the body), altruism and evolution. The second is that it cannot account for a wide range of "anomalous" phenomena, from psychic phenomena to near-death experiences and spiritual experiences. These are "rogue" phenomena that have to be denied or explained away, simply because they don't fit into the paradigm of materialism, in the same way that the existence of fossils doesn't fit into the paradigm of fundamentalist religion. Then we will look at what spirituality has to say about each of these issues, and how it can actually resolve them (that is, solve the puzzles). We will also look into the mysterious world of quantum physics, which has always highlighted the limitations of materialism – but does so even more at the present time, now that it has become clear that quantum effects take place abundantly on a macrocosmic scale and are involved in a host of biological and natural phenomena (such as photosynthesis). Finally, I will suggest that the validity of materialism is fading, and that as a culture we are moving (slowly) towards a new "post-materialist" phase.

As a result of the investigations that make up the main part of this book, two things will become clear. First, we will see just how inadequate materialism is as a way of explaining the world, and our experience of it. Second, we will see how easily – from a spiritual perspective – the "riddles" of the materialist model dissolve away. We will see that almost every phenomenon that appears "anomalous" from the perspective of materialism can be easily and elegantly explained from the perspective of panspiritism.

It's also important to point out that these issues aren't just academic. It's not just a question of me picking arguments with materialists and sceptics because I think they're wrong. As we will see in Chapter 1, the conventional materialist model has very serious consequences in terms of how we live our lives, and how we treat other species and the natural world. It leads to a devaluation of life – of our own lives, of other species' and of the Earth itself. It is essential that our culture moves beyond materialism – and towards post-materialism – as quickly as possible.

At the same time as solving many of the riddles of science, a spiritual worldview can change our relationship to the world. It can engender a reverential attitude to nature, and to life itself. It can heal us, just as it can heal the whole world.

CHAPTER 1

THE ORIGINS OF MATERIALISM: WHEN SCIENCE TURNS INTO A BELIEF SYSTEM

The materialist belief system is so pervasive and taken for granted that we may not even be aware that it exists – in the same way that for the peasants of medieval Europe, say, the belief system of Christianity was so deeply embedded into their lives that they accepted it as reality, unaware of any alternative perspectives.

When I was about 18, a friend asked me if I wanted to go to a talk about meditation at a local library. I didn't know anything about meditation, but I was curious, so I decided to go. At one point the speaker said something like, "Meditation is a way of refining your inner being. It's a way of experiencing the well-being of consciousness. Consciousness has a natural quality of well-being." At the time I had no idea what the man was talking about. I remember thinking, "'Inner being'? 'Consciousness'? What do those terms mean? Where can they be? I'm just a brain and a body. What else is there inside me?"

Once I started to meditate, I realized that I did have an inner being. I realized that there was something non-material within me – a consciousness that did have a natural quality of well-being. But this shows how deeply I had absorbed the belief system of materialism, through my education, the media and my parents and peers. I took it for granted that I was nothing more than the physical stuff of my body and brain, and that my thoughts were just projections of my brain. I took it for granted that I was nothing more than atoms and molecules.

There was no religion in my upbringing at all. That wasn't unusual – I didn't know anyone who was religious (apart from a boy in my year at school who was a Jehovah's Witness). Even my grandparents were completely non-religious. And this wasn't because they were atheists – no one I knew would have described themselves in those terms either. It was just that religion wasn't part of our lives. It wasn't a subject that anyone ever thought about or talked about. We sometimes said prayers or sang hymns in school assembly, but no one took them seriously.

Although Britain in general is a very secular country, I later learned that was particularly true of my social class. My ancestors were factory and mill workers in the northwest of England, and religion was never important to them. Factory and mill workers laboured incredibly long hours in terrible conditions, lived in poverty and often died young. They may have gone to church on Sunday mornings – often under duress, since vicars and mill owners would often round people up or punish them if they didn't go – but they probably took the services about as seriously as I took my school assemblies.

This background meant that I absorbed a materialist worldview, rather than a religious one. Without being consciously aware of it, I adopted a mechanical vision of the world and the universe. I adopted the view that the world consists of tiny particles that arrange themselves into ever-more-complex forms, eventually giving rise – through an accidental process of evolution – to living beings, and eventually to human beings. I adopted the view that the universe functions according to rigid physical laws, like a giant machine. I learned that all of the characteristics of an individual human being were passed on from their parents, in the form of tiny units called genes, which determined not only our appearance but also our behaviour. When we did our weekly religious education lessons at school, and we heard about concepts like heaven and hell and salvation and eternity, those beliefs seemed bizarre and naive, as if they belonged to a different era of human history.

I was at school in the 1970s and 1980s, and over the following decades the belief system of materialism became more pervasive. Fields such as neuroscience, psychobiology (which attempts to explain human behaviour in neurological terms) and evolutionary psychology (which suggests that present-day human traits are evolutionary adaptations from prehistory) added new perspectives to the materialist paradigm. Even more so than when I was a child, materialist assumptions permeate our educational systems, the mass media and the intelligentsia of our culture. While there might be some popular magazines or TV programmes which discuss psychic phenomena, near-death experiences or spiritual experiences, the "serious" media rarely pays attention to such concepts, except to dismiss them. To discuss them with any degree of credence would mean exposing yourself as unsophisticated and unintelligent and risking ridicule. Certainly, very few of my academic colleagues would be willing to take such "irrational" phenomena seriously. To do so would mean a loss of credibility – perhaps even a loss of career.

Recently, I met a well-known and well-respected psychologist who told me that he had always been interested in psychic phenomena and Eastern spiritual traditions, but he had never discussed them in detail in his work. He told me that in the 1980s and 1990s, when he was becoming established as an academic, it would have damaged his reputation, and prevented him from gaining a post at a university. And once he began teaching at a prestigious university, such interests would have stopped him being able to advance in his career. In other words, if he had shown his true allegiances it would have meant being excommunicated. Fortunately, the psychologist told me that – now that he had gained some status, and was nearing the end of his career – he was beginning to address these forbidden topics.

The tenets of materialism

Before we go any further, let's define exactly what materialism is. In philosophical terms, materialism is a form of monism. Here "mon" literally means "one", so we could call it "oneism" – the belief that the world consists of one fundamental or primary thing. And according to materialism, this primary thing is matter. There are no "higher" levels of reality, no different dimensions, no heaven or hell, or gods or spirits. Human beings do not have souls or spirits, and even our minds are material in the sense that they're just a product of our brains. Even the various forms of energy (such as mechanical, thermal and kinetic energy) are material in the sense that they are properties of material objects, in the same way that colour is a property of objects. Only the physical is real – the physical stuff of the world around us, and the physical stuff of our bodies.

An obvious alternative to monism is dualism – the belief that the world is made up of two fundamental things. One of these is matter and the other is a non-material quality, such as mind or perhaps soul or consciousness. According to dualists, mind or soul can't be accounted for in terms of matter – they are of a fundamentally different nature. But to materialists, there is nothing mysterious about the mind, or about life itself or even death – all can be explained in terms of the interactions of material elements, such as brain cells, molecules and atoms.

So materialism suggests that matter is the primary or fundamental substance in the world and that all phenomena (including mental phenomena) can be explained in terms of the interactions of matter. The basic reality of the world is microcosmic particles, which collect together and interact in extremely complex ways to produce everything we know. We living beings are simply agglomerations of particles. We are machine-like entities made up of tiny material building blocks consisting of different types of atoms and molecules working together to form different parts of our bodies and organize the interactions between them. Seen in this way, you could refer

12

to materialism as a "bottom-up" approach – that is, it tries to explain all human behaviour and experience in terms of biology, chemistry and physics.

These ideas might be said to be the primary assumptions of materialism, but other assumptions follow from them. Every religion has a number of basic tenets – principles that everyone who takes up the religion has to adopt. And here are what might be called "The Ten Tenets of Materialism":

- Life came into being by accident, through the interactions of certain chemicals. Once it had come into existence, it evolved from simple to more complex forms through randomly occurring genetic mutations acted on by natural selection. The driving force of evolution is competition, or the "survival of the fittest".

- Human beings are purely physical creatures, or machines. There is nothing more to us than physical stuff – that is, the atoms, molecules and cells of our bodies and brains. As a result, there is no such thing as a "soul", "spirit" or "life-force". These are superstitions that have been dispelled by science.

- Living beings consist of "selfish genes" whose goal is to replicate themselves. Human beings are merely vehicles for the propagation of our genetic material. The desire for genetic replication is the primary motivation of human behaviour.

- All mental phenomena can be explained in terms of neurological activity. Consciousness itself is generated by the brain. The billions of neurons in our brains work together – in some as yet undiscovered way – to produce our subjective feeling of being "someone" who can think and feel.

- Because consciousness is produced by the brain, and we are nothing more than physical stuff, there can't be any life after death. When my brain and body cease to function, my consciousness and identity will disappear just as the picture on a television screen disappears when the plug is pulled out.

- Human behaviour can be explained in genetic terms. Present-day human traits and characteristics exist because they had survival value for our ancestors. As a result, the genes they were related to were selected by evolution.
- As living beings we are isolated individuals, moving through space in separation to one another. I have my own body and brain, and you have yours; we can touch each other physically or communicate with one another through language, but our sense of identity – as produced by our brains – is essentially enclosed within the physical stuff of our bodies.
- The world exists "out there", separate from human beings. It is independent of us, and it would exist in the same form even if we weren't here to be aware of it.
- Our normal state of awareness is fairly objective and reliable, and shows us the world as it is. Any other states of awareness – altered or so-called "higher" states of consciousness – are hallucinations that can be explained in terms of aberrational brain activity.
- Paranormal or psi phenomena cannot be genuine because they contravene the fundamental laws of nature. For example, there is no known energy field which could link one mind to another and make telepathy possible, and no known force which could account for the ability to move objects by mental effort.

Many people pride themselves on holding these "rational" views, believing that the only alternative would be falling back into ignorance and superstition – a pre-Enlightenment, medieval view of the world based on faith and hearsay rather than on evidence. How could a rational, intelligent person believe in the possibility of life after death, or the existence of something non-material like a soul or spirit?

However, to what extent are these ten tenets actually based on evidence? To what extent are they assumptions rather than proven facts?

It is a fact that atoms and molecules exist. It is a fact that consciousness exists, and that it is associated with neurological activity. It is a fact that evolution has taken place. But it is an assumption that life can be explained wholly in terms of the action and interaction of various chemicals. It is an assumption that consciousness is produced by neurological activity (and therefore that consciousness ends with the death of the brain). It is an assumption that evolution can be explained wholly in terms of random mutations and natural selection.

And it is one of the purposes of this book to show that these assumptions may actually be false.

The cultural roots of materialism

Where did the materialist worldview come from? When did some of the basic findings of science become adapted into a belief system? And why did this belief system become so dominant? There are both cultural and psychological reasons for this, which I'll look at in turn.

Materialism didn't become dominant due to a systematic campaign of promotion or dissemination, as was the case with some of the world's most prevalent belief systems (for example, when St Paul established the basic principles of Christianity, or the Buddha established the basic principles of Buddhism). This gradual development, without any formal instigation, is probably one of the reasons why many people don't realize that materialism actually is a belief system.

There were some ancient philosophers who put forward materialist views, particularly in ancient Greece and Rome. For example, the Roman poet Lucretius wrote a poem called "De Derum Natura" ("On the Nature of Things"), which described the universe as a giant machine, and explained mental and physical phenomena in terms of tiny elementary particles (atoms). Lucretius's aim was to free Romans from superstitions, and to convince them that the world operated by chance rather than the intervention of traditional Roman gods.

In modern times, however, the foundations of materialism were established by early scientists such as René Descartes and Isaac Newton, who realized that living beings and even the whole universe itself could be understood in mechanistic terms. At the same time, such scientists were not materialists in the modern sense, since most of them were religious. Descartes was a dualist, who believed that body and soul were two different substances, while Newton saw his scientific work as an attempt to understand and explain God's creation. Newton spent much of his life writing theological works that he believed were more significant than his scientific treatises. As he wrote in his primary scientific work, the *Principia Mathematica*: "This most beautiful system of the sun, planets, and comets, could only proceed from the counsel and dominion of an intelligent and powerful Being."[1] Early astronomers had a similar attitude to Newton. For example, the investigations of the German mathematician and astronomer Johannes Kepler were motivated by his sense that God had created the universe according to geometric principles, and that human reason could uncover these.

However, in the second half of the 19th century scientific discoveries – in particular, Darwin's theory of evolution – made Christian beliefs less feasible as a way of explaining the world. It was no longer viable to believe that God had created the world, and human beings. The authority of the Bible as an explanatory text was fatally damaged. Scientists began to realize that religion wasn't even necessary to help explain the world. The new findings of science could be utilized to provide an alternative conceptual system to make sense of the world – a system that insisted that nothing existed apart from basic particles of matter, and that all phenomena could be explained in terms of the organization and the interaction of these particles. One of the most fervent late-19th-century materialists, TH Huxley, described human beings as "conscious automata"[2] with no free will. Another prominent scientist of the time, Henry Maudsley, stated that "mind is an outcome and function of matter in a certain state of organization."[3]

The First World War was also probably a significant cultural factor in the rise of materialism. The war was such a cataclysmic event – by far the most destructive and brutal war in history at that point, with 18 million people dead, millions more maimed and disabled, and all without any clear reason – that it brought about a collapse in values. It led to a distrust in abstract philosophical systems and beliefs, and a desire to pare things down to their simplest and most certain forms. It also accelerated the decline in institutional religion. The First World War seemed to offer proof of what the German philosopher Friedrich Nietzsche had proclaimed 30 years previously – that God was dead. How could a deity permit senseless destruction on such an enormous scale? How could a species that could sink to such depths of depravity and destruction possibly be made in God's image?

In the 1920s the desire to pare things down led to behaviourist psychology, which suggested that all human behaviour was simply the result of environmental influences, and that mental phenomena and consciousness itself could be disregarded because they could not be observed. In philosophy, the same impulse led to the field of logical positivism, which held that only things that could be observed and verified by the senses were meaningful, and that metaphysical statements could be disregarded because they couldn't be verified.

Shortly afterwards, the discovery of genes offered another way in which things could be pared down, and led to a new understanding of evolution (that became known as Neo-Darwinism), which in turn led to the field of evolutionary psychology. At the same time, the medical advances of the 20th century were amazingly successful, defeating illnesses that had blighted human life for thousands of years. This lent support to the idea that the human body is nothing more than a very complex machine, which can be fixed when it malfunctions. The fields of neurology and neuroscience – facilitated by brain-imaging technologies – applied this model to the brain, which was also seen as a very complex machine whose interactions could account for human experience and behaviour. All of these developments seemed to suggest that the reductionist enterprise of "paring things down" to their essential elements was valid.

As a result, materialism took hold as the dominant explanatory paradigm of our culture. Every culture has to have a metaphysical system to make sense of the world, a belief system that answers fundamental questions about human life, the world and reality itself. And since religion was no longer – for most educated people – a viable metaphysical system, materialism performed that function.

It might seem strange to use the term "metaphysical" in connection with science. Metaphysics is the area of philosophy that deals with the big questions about life and the world that most human beings wrestle with in their minds at some point – questions like, What is the nature of reality? Does life have a meaning? Is there life after death? Is the world an illusion generated by our minds? The conventional view of science is that it focuses on hard facts and experimental work rather than dealing with these kinds of questions. But the materialist outlook that many scientists adopt (often without realizing) is a metaphysical system in the sense that it has its own answers to all of the big questions, and its own perspective on the nature of reality. For example, the metaphysical system of materialism tells us that matter is the primary reality, that there is no God nor soul nor life after death, that life has no meaning except survival and reproduction, and so on.

This is the main similarity between materialism and religion – that it is a metaphysical system that explains the nature of reality. When metaphysical systems become dominant, they tend to become exclusivist and dogmatic, suppressing dissent and dismissing any evidence that seems to contradict their tenets. This has certainly been the case with religion, and to some extent – as we will see throughout this book – it is also the case with materialism.

The psychological roots of materialism

There have been times in human history when certain belief systems have taken over whole societies in an almost uncanny way, because they seemed to resonate with the zeitgeist (or spirit of the age) and satisfy the general population's psychological needs. The rise of Christianity was largely due to it becoming the "official" religion of the Roman Empire after the conversion of the emperor Constantine in the 4th century. (This is why the Pope still lives in Rome!) But the religion spread so quickly and widely that it seems clear that psychological factors were involved too. In my view this was probably due to the intense suffering and hardship of life in the medieval era, when the great majority of people lived in abject poverty and oppression, and faced constant danger of death from diseases such as smallpox, typhus and the plague. The notion that there was an almighty being watching over the world and controlling all its events became appealing to people, providing a sense of security and focus. At the same time, the idea that this life of suffering was only a brief preparation for a blissful and peaceful eternity in heaven also seemed very appealing. And in a more general sense, Christianity offered a coherent metaphysical system to explain human life and the nature of reality.

And I think the same is true of materialism. The main reason why materialism has become so popular – even more significant than the cultural factors mentioned previously – is because it satisfies deep-rooted psychological needs.

One of these needs is the same one that every metaphysical framework satisfies: a psychological need for certainty and orientation. As a belief system, materialism offers a coherent and all-encompassing explanatory framework with a credible narrative to make sense of human life. It tells us where we are, how we got here and where we're going. The explanatory power of materialism is impressive, and makes it a good replacement for religion. The answers that it offers to many of the "big questions" are clear, and they fit together elegantly and systematically. Based on the same fundamental principles, materialism answers questions about how the universe began, how life began, how it evolved, what the essence of things is (matter), why human beings behave the way we do (because of our genes and brain activity), whether there is life after death (there is none), and so on. All of this provides us with a sense of orientation. As the psychologist Erich Fromm pointed out, "man's awareness of himself as being in a strange and overpowering world" creates an intense need for a "cohesive frame of orientation" to explain the world, and reduce existential confusion and doubt.[4] This helps to give us a sense of identity too. Knowing where we are – and how we got here – contributes to our sense of who we are. Our beliefs strengthen our sense of self.

Materialism also satisfies a psychological need for control and power over the world. As the 17th-century scientist Francis Bacon remarked, knowledge is power. Feeling that we understand how the world works gives us a sense of authority and domination. Rather than being subordinate to the mysterious and chaotic forces of nature, we feel that we *over*stand (rather than just *under*stand) the world, from a position of power.

In some especially fervent materialists this psychological need for control and power manifests itself in a desire to *conquer* nature – that is, a desire to have a complete understanding of the world and the universe, to explain all their mysteries, so that we can become complete masters of creation. Essentially, this is a colonial attitude. It sees nature as uncharted territory, which it is our duty to explore and colonize. Francis Bacon explicitly compared the scientific enterprise to the colonial enterprise, believing that it was human beings' destiny – and right – to have dominion over nature. As he wrote, "Let the human race recover that right over nature which belongs to it by divine bequest."[5] (Incidentally, Bacon was also heavily involved in the colonial enterprise itself, helping to establish British colonies in Virginia at the beginning of the 17th century.)

The attitude is implicit in the way some scientists view nature as something "out there", a domain that is foreign and separate to the consciousness that observes it. Typically, when we perceive something as "other" – such as another nation, ethnic group or religion – then we also see it as an enemy to subdue and conquer. (This colonial instinct in science was especially strong a few decades ago, when such a colonization still seemed feasible, and it even seemed that we were quite close to reaching a point of complete understanding. In recent years, however, scientists have become more circumspect. More recent discoveries – such as dark energy and quantum biology – have emphasized the limitations of our understanding. In some ways, it seems that the deeper we look into reality, the more mystery we uncover.)

These psychological aspects make it clear why some fervent materialists react with such hostile scepticism to "rogue" phenomena such as near-death experiences, telepathy and precognition, dismissing out of hand compelling evidence for them. (We will look at these issues in more detail later.) Their reactions are similar to those of the Church leaders who punished early scientists like Galileo and Giordano Bruno for heresy. Like those Church leaders, they are trying to maintain a metaphysical system that satisfies their psychological needs for orientation and control. To accept the existence of phenomena which contradict the tenets of their belief system would be psychologically dangerous, threatening their identity, security and power.

Materialism and our "sleep" state

The final – and most significant – psychological reason for the success of materialism is that, as a philosophy, it corresponds very closely to our experience of the world. Or to put it another way, it is a conceptual expression of the reality we experience in a normal state of being.

In previous books such as *The Fall*, *Waking From Sleep* and *The Leap*, I have suggested that what we think of as a "normal" state of being is actually very limited and unreliable. I have even suggested that normal awareness is a kind of "sleep", which has two main characteristics.

The first characteristic is our strong sense of individuality and separateness from the world around us. Our normal experience is to feel that we are an "I" that lives inside our own mental space, with a boundary between us and the "external world". We feel that we are "in here" with the rest of reality "out there". This strong sense of individuality creates an uncomfortable sense of isolation and lack, which I think is the root of the impulse to accumulate possessions, wealth, status and power. As separate entities we feel incomplete, fragments broken off from the whole, and by accumulating wealth or power we're trying to bolster ourselves, to make ourselves feel stronger and more significant in order to compensate for our sense of lack.

This strong sense of individuality may also create a sense of otherness to our own bodies. Rather than feeling that we are our bodies, we might feel that we're just inhabiting them, as if they are merely vehicles that are carrying us around. And in many cultures throughout history, this sense of otherness to the body has led to a sense of disgust towards the body and all its functions that has manifested itself in sexual repression and asceticism.

The second main characteristic of our "sleep" state is our "desensitized" or automatic perception of the phenomenal world. The world around us is only half-real; we perceive it through a veil of familiarity, paying little attention to our everyday experiences. When we are first exposed to new experiences and environments they affect us intensely (for example, the first few days in an unfamiliar foreign country; the first few days in a new job; or the first exposure to a new smell or taste). But we quickly become habituated to them, and they lose their sensory power. The vividness of things fades away as desensitization takes place. Since most of us spend our lives in familiar surroundings, repeating experiences we have had many times before, this desensitized perception is our normal mode. We only "wake up" out of this mode under special conditions, such as when we have new experiences, travel to new environments or experience higher states of consciousness.

Most of us think of our normal state of being as a "given" and assume that the experience on the world that it gives us is true. We assume that it is the correct way to perceive the world, rather than realizing that it's just a particular vision of the world generated by our psychological structures and functioning. As mentioned, one of materialism's assumptions is that only our normal state of awareness is reliable and objective, and that any other states of awareness are aberrational (and the result of abnormal brain activity).

However, as I pointed out in *The Fall*, most cultures in human history have experienced the world in a very different way to this. There is a great deal of evidence that prehistoric human beings, and many of the world's indigenous cultures, did not experience a sense of separateness to their environment. They felt intimately bonded with the landscape, with the natural phenomena around them and with the Earth itself. At the same time, evidence suggests that prehistoric and indigenous peoples didn't experience the same desensitized vision of the world as we do. The natural world seems to have been intensely real and alive to them, full of animate and sentient phenomena. Many indigenous groups today have a strong sense that the world is pervaded with a spiritual force and that natural things are expressions of this force – as they are themselves. (More on this in the next chapter.)

We also experience a very different perception of the world in childhood. Children don't experience a sense of separation from their environment, and they also don't experience our desensitized perception. To young children the world is an incredibly real and exhilarating place, full of strangeness and wonder. They excitedly give their attention to all kinds of "mundane" and "ordinary" things that adults don't bother looking at. Our sense of separateness and desensitized perception begin to develop during late childhood and become established during adolescence or early adulthood.

But even as adults we occasionally wake up from our normal "sleep" state, when we have what I call "awakening experiences". These often occur in moments of relaxation, such as meditation or contact with nature, when the normal associational chatter of our thoughts fades away and our inner being seems to become more still and energized. Our perception becomes more intense; things become more vivid and significant, as if they've taken on a quality of "is-ness"; we also feel a strong sense of connection to our surroundings, as if we have become part of the world rather than just observers of it. As I showed in *The Leap*, it is possible to experience an ongoing, stable state of "wakefulness" in which we permanently transcend the limitations of our normal state.

It's customary for us to think that our perception of the world is more valid than that of indigenous people's or children's. We like to think that we have advanced beyond the simple animism of indigenous peoples and have a more rational understanding of the world. In a similar way, it's easy to disregard our intense childhood perception because it belongs to an earlier phase of development, which we have advanced beyond. And it is obviously true that in many ways adulthood is a more advanced psychological state than childhood – in terms of cognitive abilities, linguistic development, organizational abilities, impulse control and so on. However, in the cases of both prehistoric/indigenous human beings and childhood our development hasn't been purely positive; it has also entailed a loss. We have lost the sense of being part of the world and the sense of the aliveness and "is-ness" of the world around us.

In my view this perception of the world is the most fundamental source of materialism. Materialism is a conceptual expression of our sense of separateness and our desensitized perception. Our separateness is conceptualized into the view of ourselves as independent, objective observers of a world "out there". (As mentioned earlier, this also generates an impulse to colonize and conquer nature through *over*standing it.) Our sense of separateness is also conceptualized into the materialist view that the world consists of discrete, distinct objects which appear to exist in separation from each other, with empty space between them. On the macrocosmic level this means that the world appears to be full of inanimate and living entities (other human beings, other animals, plants, stones and so on), which are always discrete and separate. And on a microcosmic level it means that the world is full of entities such as atoms and molecules, which can co-operate and collect together but are conceived to be fundamentally discrete.

In a similar way, our desensitized perception is conceptualized into a view that the world is a fundamentally inanimate place, and that living beings are little more than chemical machines. Life is explained in terms of chemical processes, so that seemingly animate beings are simply complex arrangements of inanimate particles and atoms. And biologically inanimate phenomena – such as stones, rocks, the sky, the sun, the moon and the Earth itself – are inert objects. And between these inert objects and phenomena there is empty space, stretching around us and above us, into the sky and beyond the Earth's atmosphere.

So the materialist outlook is not an objective reality. On a fundamental level, it's just how the world appears as it is experienced through our sense of separateness and our desensitized perception.

The cultural and existential consequences of materialism

So far I've been portraying materialism in a very negative light, but surely there are some positive aspects to it?

The German philosopher Friedrich Nietzsche pointed towards some positive aspects of materialism. Living in the second half of the 19th century, Nietzsche was fervently opposed to the conventional Christianity of his time. He believed that the basis of Christianity was "disgust with life" and "hatred of 'the world', condemnation of the passions, fear of beauty and sensuality, a beyond invented the better to slander this life".[6] He believed that to reject the idea of an afterlife – which devalued this life – led to a tremendous affirmation and acceptance of this life. By rejecting the idea that there are worlds beyond this one, we love and savour this world more fully. In a similar way, Nietzsche believed that rejecting the idea of God gave us freedom, and a tremendous opportunity for self-development. God no longer stood in our way, so we were free to create ourselves anew and fulfil our potential – even to become what Nietzsche called "supermen".

The well-known British science writer and atheist Richard Dawkins has portrayed the positive side of materialism in a similar way. Despite the apparent bleakness of his mechanistic worldview, he believes that life is full of meaning and still worth living. For Dawkins, meaning comes from the very fact that we are alive at all, when the odds are so massively against any of us coming into being in the first place; as he writes stirringly, "After sleeping through a hundred million centuries we have finally opened our eyes on a sumptuous planet, sparkling with colour, bountiful with life."[7] His second source of meaning is the wonder of existence itself, the awe-inspiring complexity and intricacy of the world. Most of the time what he calls the "anaesthetic of familiarity" dulls our minds to this, but if we could look at the world with "first-time vision" we would be continually amazed by its richness and strangeness. Dawkins believes that the purpose of our lives should be to contemplate and to study this wonder, to spend our "brief time in the sun" working towards "understanding the universe and how we have come to wake up in it".[8] In these passages Dawkins strikes a tone reminiscent of existentialist philosophers such as Jean-Paul Sartre, who tell us that life is fundamentally meaningless or absurd, but that we should value our freedom.

And all of this is valid, to an extent. Research into religion and well-being has found that, while intensely religious people have the highest level of life-satisfaction, atheists have quite a high level of life-satisfaction too – higher, in fact, than moderately religious people.[9] This is probably largely because of the sense of certainty that atheism provides. Having strong beliefs definitely leads to greater well-being, regardless of the nature of those beliefs. But the well-being of atheists is probably related to some of the factors identified by Nietzsche and Dawkins – freedom from the restrictions of religion, the freedom to live by our own values, and appreciation of this existence as the only one we have.

However, the positive side of materialism is massively outweighed by its negative effects. Although some people might react in a Nietzschean way, the most natural consequences of materialism are nihilism and hedonism. Although Dawkins's celebratory attitude is inspiring, we could argue that he's not facing up to the full consequences of his own view of the world. If human beings are, as he has suggested, nothing more than "throwaway survival machines" – if our lives have no other consequence than the replication of our genes, if the universe is empty and cold and purposeless, if there's no other causal force in the universe except blind chance – if all this is true, then no amount of complexity and intricacy can really compensate us for it. To tell us to "count our blessings" and look at how intricate everything is would be like telling a prisoner in solitary confinement to feel grateful because his cell is painted in bright colours. The most honest reaction to Dawkins's view of the world – and to the worldview of materialistic science in general – would be not to bother getting out of bed in the morning, to commit suicide, to escape from the bleak reality by taking drugs or to chase after ego-gratification and sensory thrills.

And this is the bleak legacy of materialism that we live with every day – a pervasive sense of confusion and meaninglessness. In reality, materialism has created what the psychologist Viktor Frankl called an "existential vacuum"[10] – a loss of purpose and meaning.

Although Nietzsche and Sartre believed we were free to create our own values and meanings, all we have really done is turned towards hedonism and consumerism. We haven't become supermen, we've become consumers. Materialism as a metaphysical system has given rise to materialism as a *lifestyle* – that is, a lifestyle of acquisition and consumption. Believing – if only unconsciously, at the back of our minds – that this life is all there is, and that it has no meaning apart from survival and reproduction, we have developed a "devil-may-care" attitude, a sense that we may as well just enjoy ourselves as much as possible. If this world is all there is, we may as well just take as much from it as we can, without worrying about the consequences.

In a closely related way, materialism is largely responsible for the rampant individualism of modern cultures. Consumerism and hedonism naturally lend themselves to selfishness. The aim of our lives is to satisfy our own desires rather than contribute to the world or to help others. After all, there are only finite amounts of wealth, success and power to go around, so we have to be selfish and ruthless in order to grab as much of them as we can.

More perniciously, key materialist principles such as the selfish gene and competition (as the driving force of evolution) have helped to justify the worst excesses of materialism. Of course, societies were already individualistic and competitive to some degree before materialist values became dominant, but materialism made these values much more acceptable. Materialism reinforced the notion that the main goal of life is to become successful and wealthy, and made it more acceptable for us to be selfish and ruthless, and less acceptable for us to be moral and compassionate.

This highlights one of the main problems of materialism, which is that it permits and encourages some of the worst aspects of human nature. Traditional religions protect us from some of these aspects: they encourage compassion and altruism, teach us to be co-operative rather than competitive, to be moderate rather than hedonistic, and tell us that we shouldn't expect complete fulfilment in this life. But with the decline of religion and the dominance of materialism, our hedonism, consumerism and selfishness had no checks anymore and these impulses were free to express themselves. In fact, not only this – they were actually encouraged to express themselves as fully as possible. Nietzsche hoped that a materialistic outlook would free up the higher aspects of human nature, allowing us to create new meaning and live more nobly – but in actual fact it has mainly just freed up the lower aspects of human nature and encouraged us to live more meanly. It has not led to meaning, but to meanness.

The environmental consequences of materialism

The negative effects of materialism go beyond our societies and beyond us as individuals – they affect the environment too. To a large extent environmental abuse is an inevitable consequence of our "sleep" state. As noted earlier, because we can't sense the aliveness and sacredness of the natural world we don't feel respect for it, and don't feel a responsibility to take care of it. Some indigenous peoples feel that they share their identity with natural phenomena, and as a result they sense that by hurting the natural world they are hurting themselves. However, we feel that the natural world is "other" to us; we can't empathize with it, and so don't have any qualms about abusing it.

But, again, this unhealthy attitude to nature has been sanctioned and encouraged by materialism. Materialism has "proved" to us that all things – including living things – are just chemical machines. Natural phenomena are just objects whose only value is a utilitarian one. We don't feel respect or responsibility for them, we're only concerned with the use we can make of them. The Earth itself is just an insentient ball of rock, covered with some vegetation, which we think of as nothing more than a store of resources, to provide energy and produce goods. In a similar way, materialism affirms our sense that we are distinct entities, collections of atoms with a mind that's just a projection of our brains, and so separate from the natural world – and therefore entitled to conquer and colonize it.

I have called this attitude to nature "eco-psychopathology".[11] Psychopaths are people who can't feel empathy and spend their lives ruthlessly manipulating and exploiting other people in order to satisfy their desires for control and power. And that is a perfect description of our treatment of nature: a lack of empathy, with ruthless exploitation and abuse. We are eco-psychopaths, and the ultimate consequence of this psychological disorder – which is already manifesting itself – is massive damage to the ecosystems of our planet, the mass extinction of the Earth's species and possibly the extinction of the human race itself. As Chief Seattle is reported to have said in 1854: "His [the white man's] appetite will devour the Earth and leave behind only a desert."[12]

All of this makes it very clear that we need a different metaphysical system, which can provide us with a healthier and more holistic perspective, inspire us to live more meaningfully and encourage a better relationship with our planet. That alternative metaphysical system is what we're going to look at in the next chapter.

CHAPTER 2

THE SPIRITUAL
ALTERNATIVE

What if the primary reality of the universe is not matter? What if there is another quality, which is so fundamental that it actually pervades matter, and matter is actually a manifestation of it? What if this other quality also pervades living beings, and all non-living things, so that they are always interconnected?

The idea that the essence of reality is a non-material, spiritual quality is one of the oldest and most common cross-cultural concepts in the history of the world. It's an idea that almost every one of the world's indigenous cultures has developed independently, and one that each of the world's mystical or spiritual traditions has also independently incorporated. It's an idea that has been adopted by philosophers to explain problems such as consciousness and the relationship between mind and body, and one that is implied by some of the findings and concepts of quantum physics. And most importantly for my arguments in this book, it's an idea that can help to explain some of the most puzzling and controversial issues in contemporary science, psychology and philosophy.

According to this worldview, this spiritual quality is a primary aspect of reality, like elemental forces such as gravity or electromagnetism. Or perhaps this spiritual quality is even more fundamental than these forces. Perhaps it preceded the universe, and the universe – with all of its material particles and forces and laws – is an expression of it.

One thing this means is that this spiritual quality is, to use a technical term, irreducible. In other words, it can't be reduced to anything else, or explained in terms of anything else. It is simply a fundamental quality of the universe. As such, it is everywhere and in everything. It is in us, in all other living beings, in all inanimate things and in all the spaces between all things.

What term should we use for this quality? I think it's valid to describe it as a force, partly because other universal elements – gravity and electromagnetism – are seen as forces. The term "force" also suits the active and dynamic nature of the quality. (As we will see later, in Chapter 10, this quality has an innate tendency to generate greater complexity and order in material things.) So from this point on I will refer to it as "spirit-force". This term also fits well with the terms that indigenous cultures and mystical traditions throughout history (which we will examine in a moment) have used to describe the quality.

I call this perspective *panspiritism*. Literally, "pan" means "all" or "every", so panspiritism literally means "all-spirit" or "everything is spirit". This is similar to another philosophical approach, called panpsychism (which literally means "all-mind"). However, there are some significant differences. Panpsychism suggests that the most basic particles of matter have some of form of inner being, and some form of experience; it doesn't conceive of a spiritual force that pervades all things, including empty space. Panspiritism does suggest that spirit-force pervades all things, but not necessarily that it imbues them with an inner life. (In my own view – which I will describe in more detail later in this chapter – some form of awareness or sentience and experience only arises with the first simple life forms.)

Having said that, there are some clear similarities between panspiritism and panpsychism. They are both "post-materialist" approaches in the sense that they don't believe that matter is the primary reality of the world, and that mental phenomena can be reduced to brain activity. Both perspectives propose that spirit or mind is an essential aspect of the universe, and can't be explained in material terms. Both also suggest that the universe is fundamentally alive and sentient rather than mechanistic and inert.

In this chapter we're going to look at the support and evidence for this spiritual perspective. In a sense, many of the later chapters of this book provide evidence for it too. As we will see, phenomena such as altruism, telepathy and spiritual experiences offer ample support for the concept of an interconnecting spiritual force. The fact that the spiritual worldview has so much explanatory power across so many different areas is very significant. But in this chapter we're going to look at support for the idea from other sources.

Panspiritism in philosophy

In the last chapter, we saw that materialist ideas can be traced back to ancient Greek philosophy, and the same is true of panspiritism. In fact, panspiritist views were much more common in ancient Greek thought than materialist ones. The first Greek philosopher is usually considered to be Thales, who believed that "all things are full of Gods" and that "the soul is intermingled in the whole universe".[1] Another early Greek philosopher, Anaximander, used the term *apeiron* for spirit-force, which literally means "boundless" or "infinite". He described *apeiron* as the source from which all forms arise, and to which they all return. Later Greek philosophers believed that *pneuma* – literally "air", but translated as "soul", "spirit" or "mind" – was the underlying principle of the universe, pervading and penetrating everything so that all things possessed their own soul. The Stoic philosophers saw mind and matter not as two different things but as two aspects of the same underlying principle, which they called *logos*. *Logos* – sometimes translated as "God" – was therefore inherent in all material things. Other philosophers, such as Anaxagoras, used the term *nous* ("mind"), conceiving of it a single, unifying force that animated all things. And Plato, perhaps the most famous Greek philosopher of all, also expressed panspiritist views, especially in his later works. Plato used the term *anima mundi*

– the "world-soul" – and suggested that the cosmos has a soul in the same way as the body, and that everything in existence shares this soul.

Six centuries after Plato's death, and around four centuries after the demise of ancient Greek civilization, a new wave of panspiritist ideas began with the philosopher Plotinus. Little is known of Plotinus, except that he was a Greek-speaking Egyptian who spent most of his life in Alexandria and Rome. What is certain about Plotinus, however, is that he was one of the world's most profound mystic philosophers. He taught that the fundamental reality of the universe is a spiritual force that he called "The One". The One is a dynamic and powerful reservoir of spiritual force, from which all individual beings arise. It continually creates and sustains our lives, like a fountain that pours out into our individual beings. It is the central force of the universe, and as such we feel a powerful attraction to it, a longing to regain awareness of it. As Plotinus wrote: "Each being contains in itself the whole intelligible world. Therefore All is everywhere... Man as he is now has ceased to be the All. But when he ceases to be an individual, he raises himself again and penetrates the whole world."[2]

Plotinus initiated a new wave of panspiritist philosophy, usually referred to as Neoplatonism, which flourished until the middle of the first millennium AD. After this, however, there was little formal philosophical thinking in Europe until the Middle Ages. During the 16th century a new wave of philosophical speculation began, which included panspiritist ideas. The Italian philosopher Francesco Patrizi suggested in a book called *New Philosophy of the Universe*, published in 1591, that there was a soul of the universe that pervaded all things, including the human soul, so that in a sense every soul contained the whole universe. His contemporary and compatriot Giordano Bruno also believed that "in all things there is spirit, and there is not the least corpuscle that does not contain within itself some portion that may animate it".[3] One of the greatest philosophers of the 17th century, Baruch Spinoza, also expressed panspiritist ideas. Spinoza believed that there was an underlying single essence of all reality, which he referred to both as God and nature. And as with the Stoics, Spinoza believed that this quality manifested itself in both matter and mind, so that both were essentially the same.

After this, however, panspiritist ideas faded away from philosophy (although panpsychist ideas were still prevalent). One exception was the 18th-century German philosopher Johann Gottfried Herder, who used the term *Kraft* (literally, "force" or "energy") for the underlying substance of reality. He attempted to integrate the concept of *Kraft* with new forces that had recently been discovered by scientists, such as gravity, electricity, magnetism and light, by suggesting that they were all different manifestations of the underlying *Kraft* of the universe. Another notable exception is the contemporary philosopher David Chalmers, who has suggested that, rather than being produced by the brain, consciousness is a fundamental quality of the universe.

One reason why panspiritism was appealing to Greek philosophers was because it seemed to resolve one of the most problematic issues in philosophy: the relationship of the spirit or soul to the body. (This is part of the appeal of panpsychism too.) The ancient Greek philosophers expressed the problem in the phrase *ex nihilo, nihil fit*: out of nothing, comes nothing. In other words, how could the non-material soul emerge out of the material stuff of the body? Panspiritism (and panpsychism) solves this problem by suggesting that the soul was *always* in matter.

In more contemporary terms, this issue can be framed in terms of how consciousness emerges from the brain. David Chalmers refers to this as the "hard problem" – the problem of how the grey, soggy lump of matter that we call the brain gives rise to the richness and variety of our conscious experience (we will look at this in more detail in the next chapter). According to Chalmers, it is highly improbable that we will ever be able to explain consciousness in neurological terms. As a result, we should look for an alternative explanation, which is that because consciousness "does not seem to be derivable from physical laws" it should be "considered a fundamental feature, irreducible to anything more basic".[4] Chalmers points out that in the 19th century physicists realized that electromagnetic phenomena couldn't be explained in terms of present knowledge, and so they introduced the principle of electromagnetic charge as a fundamental quality of the universe. And the same should apply to consciousness. Since it cannot be explained in terms of present theories, and isn't reducible to any other quality in the universe, it should be seen as a fundamental quality. According to Chalmers, physicists will not be able to develop a coherent "theory of everything" until they take account of consciousness as a fundamental quality.

Indigenous concepts of spirit-force

However, it is important to remember that panspiritism is much more than a philosophy. More fundamentally, it is *experiential*. An all-pervading spirit-force is not just an abstract idea – it is a real quality that can be directly perceived. And the ubiquity of the direct awareness of spirit-force by human beings throughout history (and the spiritual worldviews that this awareness has given rise to) is one of the strongest pieces of evidence for panspiritism.

Almost every indigenous group in the world has a term that describes a spiritual force or power that pervades all things, and constitutes the essence of all things. In general, indigenous peoples are not (or at least were not, until colonial times) theistic – that is, they don't have concepts of personal gods who overlook the world and intervene in its affairs. (There are some peoples who have a concept of a "creator God" as a way of explaining how the world came into being, but in almost all cases once that God has done his job he steps aside and has very little to do with his creation.) In fact, the concept of "religion" had no meaning to traditional indigenous peoples, because to them there was no separation between the spiritual and the everyday world. The everyday world is pervaded by spirit, and every activity is a potentially spiritual one, since it means interacting with spirit in some form. (Incidentally, my use of both past and present tenses is deliberate here, since I'm aware that few indigenous peoples still live a traditional way of life, but that some of them still retain this perspective.)

In North America, the Tlingit of the Pacific Northwest called spirit-force *yok*; the Hopi Indians of the arid Southwest called it *maasauu*; on the Great Plains the Pawnee called it *tirawa*, the Dakota called it *taku wakan* and the Lakota called it *wakan-tanka*; while in the Northeastern woodlands the Haudenosaunee called it *orenda* and the eastern Algonquians called it *manitou*; and so on. Every North American tribe has some equivalent term. These concepts are sometimes translated as "Great Spirit", but that appears to be a western Christian interpretation (and perhaps to an extent, after cultural contact, some Indian tribes did develop a more theistic conception). As the Lakota activist Russell Means pointed out, *wakan-tanka* is more accurately translated as "The Great Mystery". It is not conceived of as a God, or supreme being, but as a divine or sacred force – one that existed before the world began, and is everywhere and in everything.[5] One of the best descriptions of "The Great Mystery" is from a Christian missionary called Reverend Stephen Riggs, who spent more than 40 years living with the Dakota in the 19th century. He described *taku wakan* as:

> supernatural and mysterious… It comprehends all mystery, secret power and divinity. Awe and reverence are its due, and it is as unlimited in manifestation as it is in idea. All life is *Wakan*; so also is everything which exhibits power, whether in action, as the winds and drifting clouds; or in passive endurance, as the boulder by the wayside. For even the commonest sticks and stones have a spiritual essence which must be reverenced as a manifestation of the all-pervading, mysterious power that fills the universe.[6]

Elsewhere in the world, the Ainu – an indigenous tribal people of Hokkaido island in northern Japan – referred to this all-pervading spiritual force as *ramut*, while in parts of New

Guinea it was called *imunu*. In Africa the Nuer people called it *kwoth* and the Mbuti call it *pepo*. The Ufaina Indians (of the Amazon Rainforest) called it *fufaka*. In parts of Polynesia – such as Hawaii, Tahiti and Melanesia – the term *mana* referred to a sacred spiritual energy that filled the whole universe and pervaded everything.

The meaning of these terms is essentially the same. In fact, they are very similar to the ancient Greek concept of *pneuma* – or Plato's idea of the *anima mundi* mentioned earlier. These terms don't refer to a deity but to an impersonal all-pervading spiritual force. This is clear from some of the translations that anthropologists used for the terms. The Scottish anthropologist Neil Gordon Munro, who was one of the first Westerners to live with the Ainu in Japan, described *ramut* as a force that is "all-pervading and indestructible" and decided that the best-possible English translation for it was "spirit-energy".[7] Similarly, the early German anthropologist R Neuhaus translated *imunu* as "soul stuff", while the British missionary JH Holmes translated *imunu* as "universal soul" and described it as "the soul of things... It was intangible, but like air, wind, it could manifest its presence."[8]

However, these concepts aren't solely confined to indigenous cultures. There are some modern, economically developed cultures that have retained concepts of spirit-force. For example, in the Shinto tradition of Japan the term *musubi* refers to the interconnecting creative spiritual force of the universe. The world is filled with *kami* – non-physical forces (or spirits) that inhabit natural phenomena, living beings and space itself, and are seen as manifestations of *musubi*. The indigenous shamanic tradition of Korea, Muism or Sinism, is similar to Shinto. The Korean term *shin* refers to spirits or divine beings (similar to *kami*), while the term *haneullim* or *hwanin* is similar to the Japanese *musubi*, referring to an all-pervading divine force or principle. Literally, *haneullim* means "source of all being". (Of course, many tribal indigenous cultures also conceive of spirits as energy forms that interact with and inhabit natural phenomena, and are also seen as manifestations of universal spirit.)

To all these peoples – and to indigenous peoples in particular – this force is not a metaphysical speculation but a tangible reality. It is not a belief but a perception. It is not an abstraction but part of their everyday experience. This spiritual perspective pervaded their lives, just as the materialist perspective pervades our lives.

The indigenous spiritual perspective

Whereas materialism sees living beings as biological machines and non-living things as inert objects, indigenous peoples' awareness of spirit-force meant that, to them, the whole world was alive. All things were animate in the sense that they were pervaded with spirit-force, and inhabited by – or associated with – individual spirits. From the materialist perspective the world consists of empty space inhabited by inert or biologically alive objects, whereas from the indigenous spiritual perspective there is no empty space because everything is filled with spirit-force and individual spirits. As the anthropologist Tim Ingold has described it, to hunter-gatherer groups the environment is "saturated with personal powers of one kind of another. It is alive."[9]

In another sense, all things are alive because they are manifestations of spirit. Spirit is the ground out of which all things grow, and in which they always remain rooted. This is a metaphor that is often used in the Indian *Upanishads* – the ancient spiritual texts that describe the world as pervaded with *brahman*, or spirit. As the *Mundaka Upanishad* describes it: "Even as a spider sends forth and draws in its thread, even as plants arise from the earth and hairs from the body of a man, even so the whole creation arises from the Eternal."[10] In a similar way, Black Elk wrote that, to American Indians, "no object is what it appears to be but is the shadow of a Reality. It is for this reason that every object is *wakan*, holy, and has a power according to the loftiness of the spiritual reality it reflects."[11] Objects are holy because they are expressions and representations of spirit. As the Dakota Ohiyesa wrote, "In the life of the Indian there was only inevitable duty – the duty of prayer – the daily recognition of the Unseen and Eternal."[12]

This sense of the aliveness and spiritual source of all things also means that indigenous peoples see all natural phenomena as interconnected. The materialist perspective sees a world made up of separate and distinct objects, whereas the indigenous spiritual perspective sees no separation, only interdependence. Everything is intimately interrelated, part of the same web of being.

This includes human beings, too, of course. In fact, this is probably one of the most significant aspects of indigenous people's relationship to the natural world: their deep sense of *bondedness* to it. This connection is so strong that many indigenous peoples feel that the land they inhabit, and the natural world in general, is part of their identity. Whereas most modern Westerners experience themselves as being "outside" nature, looking at it from a place of separation, indigenous people feel that they *are* it.

One of the most instructive – and at the same time the saddest – anecdotes of the interactions between indigenous peoples and Europeans is from a meeting between a Nez Perce chief called Tuhulkutsut and US government representatives in 1877. The representatives wanted to buy tribal land from the chief, but his connection with the land meant that he felt unable to sell. As he said, "The earth is part of my body. I belong to the land out of which I came. The earth is my mother." Of course, this meant nothing to the government representatives, one of whom replied impatiently, "Twenty times over [you] repeat that the earth is your mother... Let us hear it no more, but come to business."[13]

There is also a pertinent story from Chief Oren Lyons of the Onondaga people, who was the first young man from his tribe to attend college. On his return from college, his father asked him, "Who are you?" When Oren replied in terms of his name, his tribe and his status as a human being, his father reminded him, "Do you see that bluff over there? You are that bluff. And that giant pine on the other shore? You are that pine. And this water that supports our boat? You are this water."[14] Again, this is very reminiscent of the *Upanishads*, particularly the famous passage from the *Chandogya Upanishad*: "An invisible and subtle essence is the spirit of the whole Universe. That is Reality. That is Atman. Thou Art That."[15]

Because they felt such a sense of connection to nature, and perceived the world as fundamentally animate, indigenous peoples felt a powerful sense of kinship and respect for nature. As Chief Luther Standing Bear – one of the most acute observers of the differences between Native American Indians and Europeans – wrote, the Indian "loved the earth and all things of the earth... Kinship with all creatures of the earth, soil and water was a real and active principle".[16] This is also typified by the Lakota holy man Black Elk, who said: "Every step that we take upon the Earth should be done in a sacred manner; every step should be taken as a prayer." Of course, this is in complete contrast to the materialist perspective, which encourages an exploitative attitude to nature. And this is why indigenous peoples have always been shocked by European colonists' treatment of the natural world, seeing land that was sacred to them as nothing more than a supply of resources to be demarcated and plundered.

Perhaps most fundamentally though, the spiritual perspective of indigenous peoples meant that they did not experience the sense of meaninglessness and alienation that is associated with materialism, and the pathological behaviour that this gives rise to, such as rampant consumerism, hedonism, status-seeking and competitiveness. Their awareness of spirit-force, and their sense of kinship with other living beings and the rest of the natural world, gave indigenous peoples a sense of being a part of a greater harmony, and of being "at home" within the world. They had a sense of being supported by the world, of being nestled comfortably within nature, and of being surrounded by sacred meaning. One Native American Indian, Thomas Yellowtail, spoke of the "scared support that was always present for the traditional Indians" and described how "[W]herever you went and whatever you were doing, you were participating in sacred life and you knew who you were and carried a sense of the sacred within you. All of the forms had meaning, even the tipi and the sacred circle of the entire camp."[17] In the words of Chief Luther Standing Bear, "Earth was beautiful and we were surrounded with the blessings of the Great Mystery."[18]

If indigenous peoples, living lifestyles that had been unchanged for tens of thousands of years, can be seen as representatives of an early phase of human development, this suggests that panspiritism was the human race's most ancient worldview, and one that, until relatively recently, was completely normal and natural to human beings. It seems that, to them, spirit-force was an obvious, everyday reality – as real as the blueness of the sky or the coldness of water. To me this suggests that spirit-force is a real phenomenon, a tangible quality that can be perceived by human beings.

Spirit-force in mystical traditions

The only problem with this argument is that most modern human beings do not apparently experience spirit-force as a reality in the way that indigenous peoples do. Why should they perceive it while we don't? Surely if this quality is a reality then it would be obvious to all human beings?

However, in my view this was one of the most significant aspects of a psychological shift that our ancestors underwent thousands of years ago (that is, the event I have referred to previously as the Fall, or "ego explosion"). More specifically, it was the result of the process of desensitization described in the last chapter, when we lost the perceptual intensity of earlier human beings (and indigenous peoples).

According to my theory in *The Fall*, this was essentially a question of energy. With that psychological shift, the individual ego became much more strongly developed, which led to a new sense of individuality and separateness (and also to heightened intellectual and technological powers). And now that it was such a powerful aspect of the psyche, the ego required much more energy to function. So energy that was previously used in direct and immediate perception of the phenomenal world was now redirected to the ego. Our perception became "automatized" as a kind of energy-saving measure, to fuel the ego. This entailed a loss of the ability to perceive spirit-force in the world. It was "screened out" of our awareness as the phenomenal world became less vivid. What had previously been an obvious, everyday reality to human beings was no longer apparent to us. The world was no longer pervaded by spirit, and therefore no longer sacred. Spirit was no longer the primary reality – matter was.

However, all was not completely lost. Almost as soon as the psychological shift occurred, small groups of contemplatives around the world discovered that it was possible to temporarily undo the effects of the shift by following certain practices or ingesting certain substances. They discovered that it was possible to "de-automatize" their perceptions and temporarily wake up to a more intense reality. Some of them did this by purposely bringing about major physiological changes through fasting, sleep deprivation, taking psychedelic substances and so on. Others did it in a more stable way, by sitting down quietly and emptying their minds (in other words, by meditating).

More significantly, some contemplatives realized that it was possible to "wake up" *permanently.* Some of them began to formulate paths towards permanent wakefulness for others to follow. These became known to us as spiritual teachings and traditions such as the *Upanishads*, Taoism, Neoplatonism, the Kabbalah, Christian mysticism, Sufism and so on.

One of the most significant aspects of these spiritual traditions is that, without exception, they include concepts of a fundamental spiritual force. They each conceive of an energy or force that pervades all things and the spaces between all things, and which underlies the whole phenomenal world in such a way that all things appear to arise from it.

The ubiquity of these concepts is as remarkable as the ubiquity of concepts of spirit-force in indigenous cultures. We've already touched on one of these: the Hindu concept of *brahman*, as described in the *Upanishads*, the *Bhagavad Gita* and other spiritual texts. *Brahman* is not a theistic conception; it has no personality, no form and no control over the events of the world. *Brahman* is the "spirit supreme" that gave rise to all things in the world, and which all things retain as their essence. It is indestructible and eternal, and has natural qualities of radiance and joy, so that to become aware of *brahman* means to attain joy. And, most importantly, the *Upanishads* tell us repeatedly that *brahman* is the essence of our being, in the form of *atman*, the individual spirit. As a result, we are always essentially one with the universe. The goal of human life is to realize this oneness, and so to transcend separation, fear and even death.

In China, the concept of the Tao, or Dao, had a similar meaning. Like *brahman*, the Tao is a spiritual force that pervades the world. It is the essence of the universe, and the source from which all things emerge. The Tao is associated with balance; it maintains the order of things. At some point in the past, according to Taoist teachers, human beings fell out of harmony with the Tao and fell into self-consciousness and self-seeking (I see this as a Taoist depiction of the Fall). And now the goal of human life – in parallel with the teachings in the *Upanishads* described earlier – is to become one with the Tao again so that our lives can become its expression and we can live spontaneously and effortlessly, in harmony with nature.

In the earliest form of Buddhism – usually called Theravada Buddhism – there isn't an overt concept of spirit-force (although some scholars have suggested that *sunyata* – usually translated as "emptiness" – can be interpreted this way). However, in Mahayana Buddhism (which developed a little later than Theravada) there is the concept of *dharmakaya*, which is similar to *brahman*. *Dharmakaya* is the underlying reality of the universe, from which all things emerge and in which all things are one. As the Buddhist teacher DT Suzuki described it, *dharmakaya* has qualities of "all-embracing love and all-knowing intelligence"[19] and enlightenment means realizing the *dharmakaya* within our own being.

In contemplative traditions associated with the monotheistic religions of Judaism, Christianity and Islam, spirit-force was usually associated with God. In these traditions, God wasn't interpreted as a personal being who overlooks the world and controls its events, but as a formless, impersonal spiritual energy or force that radiates through all creation, bringing all things into oneness. It radiates through the human soul, too, so that essentially we are one with God. Christian mystics referred to spirit-force as the "Godhead" or "divine darkness". In the Jewish mysticism of the Kabbalah it was called *en sof* – literally, "without end".

There is certainly some variation among these concepts due to the concepts of the religious or metaphysical systems they were associated with. (For example, the Tao is more dynamic and tangible than *brahman*, and in monotheistic traditions spirit-force is usually seen as transcendent as well as immanent.) Nevertheless, the essential similarity of the concepts – and their similarity to indigenous people's concepts of spirit-force, and to the ancient Greek concepts of *pneuma* and *anima mundi* – is very striking. We seem to be dealing with a fundamental quality of the world that can be directly perceived. The quality may be interpreted slightly differently, from the perspective of different traditions, in the same way that, say, a landscape may be described differently by people who are looking at it from different viewpoints. In the words of the Christian and Hindu monk Bede Griffiths:

> This is the great Dao… It is the *nirguna brahman*… It is the *dharmakaya* of the Buddha, the "body of reality"… It is the One of Plotinus which is beyond the Mind (the *Nous*) and can only be known in ecstasy. In Christian terms it is the abyss of the Godhead, the "divine darkness" of Dionysus, which "exceeds all existence" and cannot be named, of which the Persons of the Godhead are the manifestations.[20]

Spirit-force outside spiritual traditions

It's also worth mentioning that concepts – and an awareness – of spirit-force are by no means confined to mystics and contemplatives associated with spiritual traditions. There has always been a very close connection between poetry and spirituality, and a great many poets – such as William

Wordsworth, Walt Whitman and DH Lawrence – were clearly aware of a spiritual force pervading the world, animating all things and bringing them into oneness. For example, in his massive autobiographical poem *The Prelude*, William Wordsworth describes how, as a young man, he could sense "the sentiment of Being spread /O'er all that moves and all that seemeth still". This also gave him a sense that everything around him was sentient, and that "the great mass [of natural things] lay embedded in some quickening soul."[21] (Wordsworth's poem "Tintern Abbey" also provides a beautiful description of this.) The great American poet Walt Whitman had an especially strong sense of spirit-force pervading all things, including his own being. For example, in one of his most beautiful short poems, "On the Beach at Night, Alone", Whitman describes his awareness of spirit flowing through all things and bringing them into oneness: "This vast similitude spans them, and always has spann'd, and shall forever span them, and compactly hold them, and enclose them."[22] (For a fuller discussion of poetry and spiritual awakening, see my book *The Leap*.)

Awakening experiences can be seen as encounters with spirit-force. As we will see in Chapter 7, awakening experiences happen most frequently to "ordinary people" who aren't affiliated with any spiritual traditions. They also usually happen spontaneously, in the midst of everyday activities and situations rather than in association with spiritual practices. In lower-intensity awakening experiences (which are by far the most common) we experience some of the effects of spirit-force rather than encountering it in a direct way. In these moments spirit-force makes the world around us more real and beautiful, gives us a sense of the interconnectedness of things, makes us feel connected to the world and gives us a sense of the radiance and harmony of things. But in high-intensity awakening experiences we may have a more immediate and powerful encounter with spirit-force, one in which we become aware of the essential oneness of all things, and of our essential oneness with the whole universe. Time and space may seem to dissolve away, leaving nothing but an all-pervading quality – which may be described in terms of energy, love or spirit. A person's sense of identity may expand massively, so that they feel that they are everything and everywhere at the same time.

In this example from my own collection a man describes an awakening experience he had two months after the birth of his son, when he was pushing a pram through the streets of his town. It was the first time he had taken his son out and he "was feeling very proud to be a dad". Then, in his words:

> I became aware of the feeling of unconditional love, not just toward my son but to everyone I was passing in the streets. It was as if I was giving it and receiving it at the same time. Then the feeling extended to inanimate objects; the path, lampposts, buildings, cars, sounds of music; everything was "made" of the same "stuff" and the only word I could find to describe it was love. Everything was made of love. I felt immersed in a sea of love where everyone and everything were made of this same "energy"; I was no longer a separate "ego" but was consumed by this energy of love. Everything became One and I was outside of time. I continued walking through a park with a very strong sense of compassion and love toward all that I encountered. The experience lasted about 20 minutes (I've since retraced my steps) and the powerful effects lasted for two or three more days. What has remained since then is an increase in empathy, tolerance, compassion and love.

One significant point here is that awakening experiences are *higher* states of consciousness. Both temporary awakening experiences and the permanent state of "wakefulness" cultivated by the adepts of spiritual traditions represent an expansion and an intensification of ordinary states of awareness. They are states in which we transcend the limitations of our normal consciousness and so become aware of qualities that are normally hidden from us. So, again, this seems to indicate that spirit-force is a fundamental reality of the world but one that is simply "screened out" of our normal awareness.

And it is also significant that when we experience wakefulness we feel the same sense of "at-home-ness" of indigenous peoples. One of the most profound effects of both temporary awakening experiences and permanent wakefulness is a falling away of the inner discord and discontent of our normal state. Instead there is a sense of ease and wholeness, an ability to live comfortably within one's own being, and in the present moment. In his poem "Pax", DH Lawrence described this sense of "at-home-ness" as being:

> Like a cat asleep on a chair
> at peace, in peace
> and at one with the master of the house, with the mistress,
> at home, at home in the house of the living,
> sleeping on the hearth, and yawning before the fire.[23]

Panspiritism in science

It's interesting to note that many of the world's greatest scientists have adopted panspiritist views too. This has particularly been true of quantum physicists. I'm not going to address the topic of quantum physics in detail here because it has its own chapter later in the book. We will see then that many of the puzzling findings of quantum physics – for example, how "individual" particles may behave as if they are twinned or entangled, and how the expectations of the observer affect the outcomes of experiments – are completely compatible with the panspiritist vision of an interconnected universe. At the very least, these findings show that the materialist view of the world is far too simplistic, and that things are much stranger than they appear to be on a macrocosmic level.

And this is presumably the reason why so many quantum physicists have advocated "post-materialist" views of reality, including panspiritism. This was especially true of many of the "founding fathers" of quantum physics, such as Werner Heisenberg, Erwin Schroedinger, Wolfgang Pauli and Max Planck (to mention just a few). Many of these physicists' metaphysical speculations are difficult to distinguish from the writings of mystics such as Plotinus or Meister Eckhart. Erwin Schroedinger – most famous for the "Schroedinger's cat" thought experiment, which shows that the state of a phenomenon is indeterminate until it is observed – saw his scientific investigations as a way of approaching the essential oneness of the universe, of which our individual consciousness was the manifestation. As he wrote in his book *My View of the World*: "Inconceivable as it seems to ordinary reason, you – and all other conscious beings as such – are all in all. Hence this life of yours which you are living is not merely a piece of the entire existence, but is in a certain sense the whole."[24]

It may seem surprising to hear scientists speaking in such spiritual terms, but this is only because we've come to associate science with scient*ism*. These views illustrate that science doesn't have to be imbued with materialism, and that there isn't necessarily a conflict between science and spirituality. In quantum physics – which can be considered as the most fundamental of sciences, since it deals with the most microcosmic aspects of reality, which inform all others – science and spirituality become reconciled.

Can spirit-force be detected?

You might argue that if all of that is true – that is, if spirit-force is a real quality that can be sensed by human beings – why isn't it an established scientific concept? Why are scientists seemingly unable to detect it?

One important point here is that spirit-force – or universal consciousness, if you prefer – is non-physical. It isn't made of atoms and molecules, and so it can't be directly observed or detected. You can't take out a telescope and expect to see spirit-force pervading space; you can't take out a microscope and expect to see it pervading atoms. That would be the same as doing a brain scan and expecting to "see" consciousness. (In fact, since our own consciousness is a canalization of universal consciousness, this literally is the same thing.)

Another reason why it's impossible to observe or measure spirit-force in the same way as physical forces or objects is because *we are it*. It is not outside us. There is no possibility of us getting outside ourselves to measure it. To detect something means to look at it externally, as an object. But we can never look externally at spirit-force, or consciousness. We are always looking with it. When we look, it is looking through us.

In fact, there are many scientific phenomena whose existence is taken for granted even though they are non-physical and can't be directly detected or measured. Until gravitational waves were observed for the first time in 2015, there was no evidence for the existence of gravity apart from the observation of its effects. No one has ever seen quantum particles such as quarks and photons, but their existence is assumed based on their effects. In the same way, no one has directly detected dark matter – its existence is inferred because of the gravitational effects it appears to have on galaxies and galaxy clusters. You could compare all of these phenomena to the wind – no one has ever seen the wind, but we know it exists because we can see its effects on our surroundings and our own bodies.

In the same way, I would argue that we can assume the existence of spirit-force – or universal consciousness – because we can sense its effects. Its effects can arguably be detected in quantum physics – in terms of the entanglement of particles and the dissolving of the duality between the observer and the observed. Its effects can be sensed on a psychological or spiritual level – for example, in our feelings of empathy towards other people, or other living beings, and our sense of connection to nature. As previously noted, in higher states of consciousness, or awakening experiences, we can sometimes perceive qualities of spirit-force, such as interconnection, radiance and harmony. In awakening experiences it is also possible for us to directly perceive spirit-force in the world. As we have seen, to indigenous peoples this was completely normal. So in this sense, even though it can't be measured or detected, spirit-force *is* tangible.

My own perspective

As I noted earlier, one of the main differences between panspiritism and panpsychism is that the former doesn't suggest all things have their own mind, or inner being, and therefore their own experience. My view is that although spirit-force is in all things, not all things have their individual spirit. Or to put it more clearly, although consciousness is in all things, not all things are conscious. That is, not all things have their own individualized consciousness. Only the structures – beginning with cells – that have the necessary complexity and organizational form to receive and canalize consciousness are individually conscious and individually alive. In my view, this is the primary function of cells: to facilitate the canalization of spirit-force into individual beings. A cell acts as a "receiver" of consciousness, so that even an amoeba has its own very rudimentary kind of psyche, and is therefore individually alive. And as living beings become more complex – as their cells increase in number and become more intricately organized – they become capable of "receiving" more consciousness. The raw essence of consciousness is channelled more powerfully through them, and they become more intensely alive, with more autonomy, more freedom and more intense awareness of reality. That's why human beings, with our incredibly complex and intricate brains, are one of the most conscious beings (perhaps alongside dolphins and whales) that evolution has yet developed.

However, the simplest forms of matter, which do not have cells, are not capable of canalizing consciousness, and so they are not individually conscious or alive. Simple forms of matter do not have an interior, and are not capable of experience or sensation. These only emerge at the cellular level and above.

In a sense, all things are alive, as many indigenous peoples believe, since they are all pervaded by consciousness, or spirit-force. But there is a difference in the way that rocks and rivers are alive and the way that an insect or even an amoeba is alive. Rocks and rivers do not have their own psyche, and are therefore not individually conscious. Consciousness pervades them, but they aren't conscious themselves. They can't be, because they don't have any cells – let alone brains or nervous systems – to canalize consciousness.

So there is a distinction between individual conscious beings and consciousness as a whole. Individual beings still exist, with cells or brains which canalize all-pervading spiritual force into them to different degrees. There is a distinction between material things, which are just pervaded with spirit-force (without having their own interior minds), and living things, which are both pervaded with spirit-force and also have their own interior mind or consciousness.

Spirit-force therefore manifests itself in two ways: as matter and mind. You could say that matter is the external manifestation of spirit, while mind is its internal manifestation. All matter is the manifestation of spirit, but some complex forms of matter also have spirit as an internal quality. Another way of looking at this is to think in terms of two different stages. The first stage, at the beginning of the universe 13.7 billion years ago, was the emergence of matter out of spirit. The second stage, which took place around nine billion years later, was the emergence of mind within matter, which began with the first simple life forms. (At least, this process began on Earth nine billion years later – for all we know, it may have happened earlier on other planets.)

In other words, my view is similar to that of the Greek Stoic philosophers and Spinoza, as discussed near the beginning of this chapter – that the essence of reality is a quality that manifests itself in both mental and physical terms. Spirit precedes both mind and matter, and is the source of both.

It goes without saying that panspiritism is a much healthier perspective than materialism. It's healthier for us as individuals, for the human race as a species and for the whole of our planet. Whereas materialism stems from – and further encourages – anxiety and accumulation, the panspiritist perspective lends itself to ease and contentment. Whereas materialism encourages individualism and competitiveness, panspiritism leads to empathy and altruism. Whereas materialism promotes exploitation and domination of the natural world, panspiritism engenders respect and harmony. Whereas materialism can only lead to the devastation of our planet, and perhaps even to our extinction as a species, a panspiritist perspective is perfectly sustainable, and offers us a harmonious future.

However, over the next few chapters of this book the main aspect of panspiritism we're going to focus on is its explanatory power. We're now going to begin examining some of the puzzling phenomena which make little sense from a materialist perspective but which are easily comprehensible in terms of panspiritism. And the first of these puzzling phenomena follows on logically from this discussion: consciousness.

CHAPTER 3
THE RIDDLE
OF CONSCIOUSNESS

Francis Crick was one of the most eminent scientists of the 20th century. In 1953 he helped, along with James Watson, to "break" the genetic code (by discovering the structure of the DNA molecule). Later in his scientific career, Crick decided to turn his attention to what he saw as the biggest remaining problem in science: consciousness. He decided he was going to solve the riddle of how the brain produces our "inner life" of thoughts and sensations.

Crick fully expected the riddle to be solved within a few years, with the help of the latest brain-scanning and imaging technology. The issue seemed straightforward: human beings experience consciousness, and consciousness is produced by the brain. After all, isn't it clear that when the brain is damaged, consciousness is impaired? And isn't it clear that when the brain stops functioning – at the point of death – consciousness stops too? As Crick put it, graphically: "You, your joys and your sorrows, your memories and your ambitions, your sense of personal identity and free will, are in fact no more than the behavior of a vast assembly of nerve cells and their associated molecules."[1] His task was therefore clear: to investigate exactly how these nerve cells and molecules gave rise to our conscious experience.

Unfortunately, consciousness proved to be much harder to "crack" than the genetic code. Working together with a young researcher called Christof Koch, Crick devoted the last two decades of his life to the riddle of consciousness but made frustratingly little progress. He made a number of suggestions – for example, that consciousness was related to the brain's visual cortex, to short-term memory or "some form of serial attentional mechanism"[2] – but none of these were confirmed by evidence.

Interestingly, although Crick never gave up his faith in a materialistic explanation of consciousness, his co-researcher did. Christof Koch eventually came to doubt the basic assumption of their work: that consciousness can be explained in terms of brain activity. He began to investigate alternative ways of explaining consciousness, and adopted a panpsychist perspective. (We will look at his views in a little more detail later in this chapter.)

In this chapter, we will examine why Crick and many other scientists have made so little progress in attempting to explain consciousness in terms of brain activity. And we will see how, from a spiritual perspective, consciousness begins to seem much less problematic. In fact, we will see that panspiritism offers a sensible solution to the "riddle of consciousness".

Defining consciousness

First of all, let's define exactly what we mean by consciousness. Consciousness is one of those words – like "spiritual" or "spirituality" – that is used in so many different contexts with so many different connotations that it's difficult to define. Even scholars and theorists of consciousness sometimes use the term with slightly different meanings. So I think it's best to use quite a broad definition. At my university, I teach a module on Consciousness Studies to first-year undergraduates, most of whom are 18 years old. Other academics sometimes say to me, "How can you possibly teach consciousness to 18-year-olds?" But I find that once we agree on a definition of what consciousness is, then they are surprisingly clear about the topic and become very engaged with it.

So what is consciousness? One reason why it's tricky to define is because it's *us*. We are consciousness, so it's difficult to step outside ourselves and observe it as if it's something "other" to us. In view of this, the best way to understand consciousness is in terms of experience rather than definition. That's why, in the very first session of our Consciousness Studies module, I guide the students through an exercise – a sort of meditation. In fact, I'll give you – the reader – the same exercise now:

Close your eyes and observe your own inner experiences. Watch your thoughts pass by, as if you're sitting on a riverbank watching a river flow by. These could be thoughts about what happened earlier today, about what might be happening later today, about the other people around you, about this book and so on. The important thing is just to watch the thoughts arise, manifest themselves and fade away.

In the same way, be aware of any sensations inside you – for example, any feelings of discomfort or irritation or tiredness. Again, just be aware of those feelings, with the sense that you are apart from them, as an observer. Also, be aware of the chair you're sitting on, of the sensations of your back against it, and your bottom upon it. Be aware of your feet against the floor.

Now – still with your eyes closed – try to sense that part of you that is aware of your thoughts and sensations. Since you are watching your thoughts pass by, there is a part of your consciousness that is apart from your thoughts – a watcher, or observer. In metaphorical terms, this is the part of you that is sitting on the riverbank watching the river of thoughts flow by. This is your sense of "I". After a while, you may gain a sense of the distance between this "I" and your thoughts. You may also be aware of how your thoughts try to pull you away from this place of observation, how they immerse your attention, as if the river is trying to carry you away.

Finally, let's bring consciousness outside ourselves. Still with your eyes closed, be aware of the sounds in the room, and outside it. Be aware of any aromas around you. Then touch some of the objects around you. Then open your eyes and look at the objects and people and different phenomena around you. Be aware of your surroundings through all your senses.

This exercise illustrates three different aspects of consciousness. The first aspect is our inner experience of thoughts and sensations. Philosophers of consciousness call these inner experiences "qualia". In the singular, a quale is a unit of consciousness experience. A quale can be the taste of a tomato, a sensation of pain as you accidentally touch a red-hot stove or an anxious thought about a future event.

The second part of the exercise illustrates that we appear to have a centre of consciousness, a sense of "I" with which we are aware of our own experience. This means that we don't just have experience, we're also aware of it. In other words, this is the part of us that is self-conscious. It watches our thoughts, observes our interactions with other people, commentates on and criticizes our behaviour, and so on. This self-conscious observer is a second aspect of consciousness.

The third part of the exercise illustrates that consciousness includes our awareness of our surroundings. This awareness works through our senses and puts us into contact with the world outside us. This is the third aspect of consciousness.

One of the interesting questions about consciousness is whether it's just a human phenomenon or is shared by other animals? The three-aspects definition perhaps throws some light on this. In terms of the first aspect – inner experience – the philosopher Descartes believed that only human beings have minds (or souls), and that animals are just automata. But most modern philosophers would be more circumspect, since it is obviously impossible to know whether animals have any subjective awareness – or qualia – or not. At the same time, it seems apparent that many animals have the capacity to feel pain, fear and even sadness (such as when elephants, apes and horses appear to grieve for deceased relatives). This obviously suggests some degree of inner experience.

In terms of the second aspect, there is evidence that some animals possess a degree of self-consciousness. A number of animals – including chimpanzees, bonobos, elephants and even Eurasian magpies – have passed the "mirror self-recognition test". When spots are placed on their faces, and they are put in front of a mirror, they react by touching the spots or trying to rub them off, as a human being would do. A variation of the mirror test was attempted with dolphins. A researcher marked parts of their body with ink, and they frequently turned those areas to the mirror, suggesting that they were making a conscious effort to see them.[3] However, most animals don't pass the mirror self-recognition test, which suggests that most don't have self-awareness.

In terms of the third aspect, we can safely say that all animals have some degree of awareness of their surroundings. Even a single-celled amoeba will move towards light and sources of food, showing awareness of its surroundings. And the more physically complex animals become, the more they display awareness of their surroundings. So in this sense, the important question is not whether animals are conscious, but *how* conscious they are.

The brain as the source of consciousness

Francis Crick's belief that the consciousness was the last remaining big question in science helps to explain its popularity over the last 30 years or so. According to this narrative, we have now reached the point where we largely understand problems like evolution, the nature of life and the origins of the universe, so it is time for us to turn our attention inside and solve the problem of consciousness. Of course, this attitude was largely based on false confidence – it is very debatable that we do largely understand the above phenomena, particularly in view of more recent findings. For example, the discovery of dark energy in the 1990s (although its existence has still not been directly detected) – together with the older theory of dark matter – has shown that there is a great deal about the universe that we don't understand. Likewise, the mapping of the "human genome" has shown that we understand much less about the genetic basis of life than we presumed. (We will discuss this in more detail in the next chapter.)

And, as Francis Crick found, the belief that we would be able to solve the mystery of consciousness was also based on false confidence. After decades of intensive research and theorizing, very little progress has been made in understanding how the neuronal networks of the brain relate to consciousness. Many suggestions have been made, besides those of Crick. For example, the Scottish philosopher Donald MacKay suggested that consciousness is related to interactions between the cortical layer and other deeper layers of the brain; the neuroscientist Rodney Cotterill has suggested that the site of consciousness is the anterior cingulate, while VS Ramachandran – one of the most eminent neuroscientists of all – has suggested that the "circuitry" of consciousness lies in the temporal lobes as well as a part of the frontal lobes called the cingulate gyrus.[4]

The wildly varying nature of these suggestions tells its own story. When so little consensus exists in explanations, it suggests that the causal assumption underlying the explanations (in this case that the brain produces consciousness) is doubtful. In fact, the idea that consciousness stems from one particular area of the brain has now been largely discarded by theorists, in favour of the view that consciousness is in some way generated by the brain as a whole. As the neuroscientist Giulio Tononi has put it, "Consciousness is associated with a distributed neural system: there is no single area where it all comes together."[5] But still no one has put forward any viable theory of how the whole brain may produce consciousness.[6]

There are other difficulties too. As Tononi has also pointed out, brain cells fire almost as much in deep sleep as they do in the wakeful state, despite the lack of (or at least a lower level of) consciousness in the former. They also fire to a high degree in epileptic absence seizures (when a person blanks out) even though consciousness is lost. In certain parts of the brain – such as the thalamocortical system – you can identify some neurons that correlate with conscious experience, while other neurons do not seem to have any effect on it. Why should consciousness correlate with some neurons but not others? All of this suggests a lack of a direct and reliable relationship between brain activity and conscious experience.

However, there is an even bigger issue: many philosophers have suggested that the very assumption that the brain produces consciousness should be abandoned. If you held a brain in your hand, you would find it to be a soggy clump of grey matter, a bit like putty, and about as heavy as a bag of flour. How is it possible that this grey soggy stuff can give rise to the richness and depth of your conscious experience? This presumption is a "category error". The physical matter of the brain – no matter how complicated the interactions between the cells are – belongs to one category of substance, and the non-physical qualia of conscious experiences belong to another, so how can the latter be explained in terms of the former? As the philosopher Colin McGinn has put it, to say that the brain produces consciousness is like saying that water can turn into wine.

Some philosophers have suggested that consciousness is an "emergent" property, which naturally arises once matter reaches a certain level of complexity. However, this is just a description rather than an explanation. Since no one has been able to explain how consciousness might emerge from matter, it is just a restating of the problem. And in any case, when a property emerges from the most basic components of a system, that property is normally inherent in those components and can be deduced from them. But there is nothing about conscious experience that is relatable to the physical stuff of the brain. At the most microcosmic level, the brain consists of sub-atomic particles, which have qualities like mass, spin and charge. There is nothing about these qualities that relates to the qualities associated with consciousness, such as thought, taste, pain or anxiety. As Colin McGinn states graphically, "You might as well assert that numbers emerge from biscuits or ethics from rhubarb."[7]

As noted in the last chapter, the Australian philosopher David Chalmers has referred to this as the "hard problem". As Chalmers sees it, there are some aspects of the relationship between cognitive activity and brain activity that psychologists and neuroscientists understand fairly well. For example, we have a fairly good idea of the brain functions involved in memory, attention and information processing. But these are just – in Chalmers's terminology – the "easy problems". The problem of how the brain might give rise to consciousness is on a completely different scale. The "hard problem" may not be soluble at all.

This has also been described as the "explanatory gap". Even if we did somehow manage to precisely identify the neural networks associated with consciousness, what would this tell us? There would still be a gulf between the physical stuff of the brain and the richness of conscious experience. (Essentially, this is the same problem that, as we saw in the last chapter, was expressed by Greek philosophers as *ex nihilo, nihil fit* – out of nothing, comes nothing.) As Christof Koch, explaining why he came to doubt neurological explanations of consciousness, has put it: "[the] emergence of subjective feelings from physical stuff appears inconceivable... The phenomenal hails from a kingdom other than the physical and is subject to different laws. I see no way for the divide between unconscious and conscious states to be bridged by bigger brains or more complex neurons."[8] Koch realized that panpsychism offered a way to transcend this problem, and he now believes that, rather than being produced by the circuitry of the brain, consciousness is "inherent in the design of the universe".

David Chalmers has illustrated this with his concept of a "zombie". Imagine that there is a zombie version of you that looks exactly like you, speaks exactly like you and behaves exactly like you, and is only different from you in that it doesn't have conscious experience. It is exactly the same as you except that there is no self-reflecting, sensation-experiencing self inside. There is no one there to think about what you're doing, commentate on your experience or make plans or decisions. Chalmers believes that, hypothetically, this zombie version of you could survive in the world. You could function perfectly well in the world without conscious experience. And there is nothing about your physical form that necessitates conscious experience. This means that consciousness is "something extra", something beyond the physical stuff of our brains and bodies that can't be reduced to them.

Consciousness as an illusion

Another possible approach to consciousness is to say that it doesn't need to be explained because it's an illusion. The most well-known advocate of this approach is the American philosopher Daniel Dennett. He believes that we don't need to explain how the physical stuff of the brain gives rise to consciousness, because it doesn't. In other words, there is no "hard problem". Dennett's response to Chalmers's concept of the zombie is to say that we are *all* zombies. None of us are really conscious, even though we convince ourselves that we are.

What Dennett tries to do is not, therefore, to explain how consciousness arises, but to try to show how the illusion of consciousness arises. He suggests that this illusion is closely related to the illusion of self. In the exercise I described earlier, the self is the aspect of consciousness that observes our own mental processes, and "looks out" at the world. Dennett describes this as the illusion of the "Cartesian Theatre", based on Descartes' famous phrase, "I think, therefore I am". It feels as if we are sitting in a theatre, watching our thoughts pass by, but in reality no one is there. According to Dennett, there are just mental processes, streams of thoughts, sensations and perceptions passing through our brains, without a central place where all of these phenomena are organized. People only believe they are conscious when they look inside and take a snapshot of these processes. But until that point there was nothing in consciousness.

However, there is a basic absurdity in the argument that consciousness is an illusion. The argument can only be made via the consciousness of individual human beings, such as Daniel Dennett. And these individuals are obviously assuming that their own consciousness is authentic and reliable – otherwise they would not bother to state its observations. If they really believed that their own consciousness didn't exist, then surely they would not trust its insights and ideas? Dennett presupposes that there is a reliable observer inside him who is able to pass judgement on consciousness – and that very presumption contradicts his own arguments, since this observer is the very thing whose existence he is trying to disprove. Would you trust the evidence of a witness who has been shown to be unreliable – indeed, a witness who you yourself have concluded does not exist? In other words, the argument is caught in a loop. Consciousness cannot prove that it does not exist.

The issue becomes even more absurd when we consider that, in order to argue that consciousness doesn't exist, Dennett collects many examples of experiments which show how unreliable human awareness can be, how we often misperceive situations and make assumptions that turn out to be false. All of these experiments were undertaken by human beings who believed they were conscious. But if the individual consciousnesses which conducted these tests and described these findings are actually illusory, why should we trust their findings?

Related to this, there is a problem of subject/object confusion. Dennett attempts to examine consciousness from the outside. He treats it like a botanist examining a flower, as an object to scrutinize and categorize. But, of course, with consciousness, the subject *is* the object. You are consciousness. So it is fallacious to examine it as if it is something "other". Again, you are caught in a loop. You can't get outside consciousness. And so any "objective" pronouncements you make about it are fallacious from the start.

In fact, what Dennett does is to simply ignore the subjective aspect of consciousness, including his own subjectivity. Like behaviourists in psychology, he believes that subjectivity can just be disregarded. But the very idea that consciousness is an illusion presupposes that there is someone *to whom* it is an illusion. And that someone is the human subject itself.

All of this shows how problematic and bizarrely counterintuitive it is to argue that consciousness is an illusion.

The spiritual alternative

Some philosophers believe that because of the difficulties of explaining consciousness in terms of the brain – and the absurdity of pretending that it's an illusion – we shouldn't expect to understand it. This is what is known as the "mysterian" position, and in some ways it makes sense. (In fact, I will make a similar argument in the last chapter of this book, although not specifically about consciousness.) The human intellect is limited; there are surely some things that are beyond our understanding, some riddles that we will never be able to solve. And the riddle of consciousness is especially problematic, since – as I've just noted – we are consciousness, and so it's impossible for us to investigate it with clarity or objectivity. Because of this, we may well be – as the philosopher Colin McGinn has put it – "cognitively closed" to the problem of consciousness. However, I believe that we can make some sense of the riddle of consciousness if we look at it from the perspective of panspiritism.

From the spiritual point of view, consciousness does not emerge from complex arrangements of material particles; it isn't located in certain areas of the brain, or produced by certain types of brain activity. Consciousness doesn't emerge from matter because it has *always* been in matter. Consciousness is a fundamental quality that exists everywhere and in everything.

From the panspiritist perspective, the brain does not produce consciousness but rather it acts a kind of receiver, which transmits and canalizes universal consciousness (or spirit-force, which is equivalent to it) into our own being. Via the brain (not just the human brain, but that of every other animal), the raw essence of universal consciousness is canalized into our own individual consciousness. And because the human brain is so large and complex it is able to receive and canalize consciousness in a very intense and intricate way, so that we are (probably) more intensely and more expansively conscious than most other animals. As the philosopher Robert Forman has put it:

> Consciousness is more like a field than a localized point, a field which transcends the body and yet somehow interacts with it... Brain cells may receive, guide, arbitrate, or canalize an awareness which is somehow transcendental to them. The brain may be more like a receiver or transformer for the field of awareness than its generator.[9]

As we saw at the beginning of this chapter, one of the most obvious reasons for assuming that the brain produces consciousness is that consciousness may be impaired or altered if the brain is damaged. And when brain functioning is altered to some degree – for example, by drugs – then consciousness is usually affected. However, this doesn't invalidate the spiritual explanation of consciousness. Even if the brain doesn't produce consciousness, but rather receives and transmits it, any damage or alteration will have an equally significant effect. A radio doesn't produce the music that comes through it, it just receives and transmits it; nevertheless, if the radio is damaged, then its ability to transmit the music will be impaired. And if someone changes the tone control of the radio (which is analogous with taking drugs) or tampers with its inner circuitry, then its output will obviously be affected.

Panspiritism also fits well with neuroscientists' assumption that consciousness is somehow associated with the brain as a whole (despite not being clear about the processes involved) rather than located in one particular part or pattern of neurological activity. If the brain's role is not to produce consciousness but to receive and transmit it, then we would fully expect it to be widely distributed in this way.

Consciousness does not depend on any particular part of the brain; the brain's "receiving and transmitting" role depends on it functioning as an integrated, interrelated whole.

It might be argued that panspiritism doesn't really solve the riddle of consciousness, because it doesn't explain where consciousness came from in the first place. But in a sense it doesn't need to do this. Consciousness doesn't come from anywhere – it just is. Physicists don't feel it necessary to try to explain where electromagnetism, mass or gravity come from – they are just built into the universe. And the same could be true of consciousness, or spirit-force. (In fact, as noted in the last chapter, consciousness may be even more fundamental that the above forces, if we assume that it actually *preceded and gave rise* to the universe.)

In a similar way, you could argue that panspiritism cannot tell us how the brain receives and transmits consciousness. It cannot identify the processes involved, just as materialists cannot identify the processes by which the brain might give rise to consciousness. This is true, of course. It could be that we will never know this – perhaps this is one sense in which the "mysterian" argument holds true, in that there are some things that our limited intellect and awareness will never be able to understand.

Nevertheless, on a theoretical level, the panspiritist argument seems very viable as an alternative to materialist explanations of consciousness. And when you take into account the wide range of other evidence for panspiritism – all of which we will examine in this book – the theory seems very convincing as an explanation of consciousness.

According to panspiritism, it isn't just a question of us having consciousness, but of us *being* consciousness. And it isn't a question of us being *individually* conscious, because we share the same consciousness. This means that we are essentially one – part of a greater unity rather that separate individuals. As we will see later, this oneness is the source of altruism and can help to explain some types of psychic experiences. It is also experienced directly in some spiritual or mystical experiences.

CHAPTER 4

THE PRIMACY OF MIND: PUZZLES OF THE MIND AND BRAIN

Some time ago I was listening to a radio programme called *All in the Mind* on BBC Radio 4. *All in the Mind* is a series about psychology, and I normally enjoy it. But this particular programme left me shaking my head. It was about memory, and included an interview with a neuroscientist who explained his theory that memory works through the hippocampus, which is where he believed memories are stored. The hippocampus had limited storage space, which is why memories are limited, and why we sometimes forget things. When new memories come in, older (or less important) ones are pushed out. He also suggested that this is why absorbing a lot of information can make us feel tired. The hippocampus can become overloaded, with too much information stuffed into it, and we need to rest and sleep in order to allow it to process the information and recharge itself.

Let's stop for a moment to consider how absurd this discussion really is. The hippocampus lies roughly at the centre of the brain, under the cerebral cortex. It looks rather like a seahorse, which is what gives it its name. Imagine if we were able to see inside the hippocampus of a student while they were revising for their exams, looking through textbooks and making notes, trying to cram as much information as they can into their mind. Would we see some little, physically real, things called "memories", jostling around and pushing each other in and out? Would we see a small pile of these things steadily growing in the hippocampus, until it was quite sizeable and the student finally became tired and fell asleep?

Of course we wouldn't. The hippocampus does not actually have any memories inside it. All you would see inside the student's hippocampus – or anyone else's, even when they're not exercising their memory – is a few centimetres' worth of soggy grey tissue. If you looked more microscopically, you would find millions of brain cells. But you would not find any memories – simply because memories are mental phenomena, not physical entities.

This is not to say that the hippocampus is not involved in memory. It obviously is – for example, we know that it becomes damaged in Alzheimer's patients, leading to loss of memory. But to say that it is involved in memory is not the same as saying that it stores memories. (On another point, the idea that memories are stored in the hippocampus creates the impression that is the only source of memory. But other research suggests that memory is not associated wholly with any particular place, but spread across many different areas of the brain. This applies to other functions too, just as it does to consciousness itself.)

This illustrates an approach that is very common among neuroscientists and scientists in general: treating mental phenomena as if they are neurological in nature – that is, treating the mind as if it is nothing more than brain. According to materialism, the mind cannot exist as a thing in its own right because it has no physical reality. Only the brain is physically real. And so mental activity must be a product of – or be equivalent to – brain activity. All mental states can be reduced to brain states. If we have psychological problems, they are due to neurological imbalances or malfunctions, and can be "corrected" by medication. If we have anomalous experiences – such as higher states of consciousness, out-of-body experiences or near-death experiences – they are due to aberrational brain activity. Such experiences have no reality in themselves, but are just brain-created illusions. As the philosopher Daniel Robinson has put it: "All mental states, events, and processes originate in the states, events, and processes of the body and, more specifically, of the brain."[1]

This attitude is so prevalent, and so embedded in our culture, that it's often reflected in the language that people use to talk about psychological issues. Neurological terms are used to describe psychological phenomena, as if they are the same thing. A person who is suffering a mental problem might say that their "brain is all messed up" or that they need to "get their brain sorted out". But, of course, they are actually talking about the mind, not the brain.

Neuromania

At the present time in our culture, there is so much enthusiasm for explaining human behaviour in neurological terms that some observers have suggested that we are afflicted with "neuromania". One manifestation of this is several new areas of academic study that attempt to incorporate neurology, such as neuroeconomics (focusing on the neurological correlates of economic decision-making), neuroaesthetics (focusing on the neurology of aesthetic experiences) and neuro-literary criticism (studying how reading literature affects the brain). The assumption behind these fields – and behind neuromania in general – is that all human experience is rooted in brain activity, and that by understanding – and altering – brain activity, we can change human behaviour.

Until the late-1990s, it was common to attempt to explain human behaviour in terms of genes. In the popular media, it was common to speak of a "gene for" various types of human behaviour – a gene for alcoholism, depression, criminality, homosexuality, a "shopping gene" for women and so on. Around that time, in 2000, geneticists were in the process of mapping the "human genome", in the hope that the genes responsible for the whole spectrum of human experience would be identified. (Sometimes the genome was referred to as "the book of life".) It was hoped that this would lead to a revolution in our understanding of everything from disease to human consciousness. However, the Human Genome Project was something of a disappointment. It revealed that genes are much less significant than scientists thought, and seemed to uncover more questions than it answered. For example, it found that human beings have far fewer genes than expected – only around 21,000, which is around 10,000 less than a tomato. We also share many of our genes with other life forms – for example, one study found that we share 97.5 per cent of the same working DNA with mice.[2] This makes it difficult to explain the amazing complexity of human beings in genetic terms, or the difference between us and other species.

Another puzzling finding of the Human Genome Project was that some inheritable characteristics – such as height – are only very slightly related to genes. In addition, although the project was predicted to bring about a revolution in healthcare, it was found that faulty genes have a less significant role than expected in predisposing human beings to disease.[3] As a result of these findings, the "genes for" discourse has begun to disappear. It seems that there actually aren't any "genes for" after all.

This is probably why neuroscience has become so popular now. It has replaced genetics as our primary explanatory tool. The emphasis has shifted away from the genome and up to the human brain. It is now "neuronal circuits" that are responsible for everything, rather than genes. Similarly to the earlier "genes for" approach, some neuroscientists and popular commentators claim to have identified the brain activity – or the parts of the brain – associated with terrorism, creativity, aesthetic appreciation, political affiliation and a host of other characteristics.

In fact, in recent years, some neuroscientists have begun to speak of mapping the "connectome" (sometimes called the Connectome Project) – that is, mapping the complete circuitry of the human brain in the hope that this will provide us with an understanding of human behaviour and experience, and the origins of "brain disorders" such as dyslexia, schizophrenia and Alzheimer's. This enterprise has replaced the earlier enterprise of mapping the genome, and it springs from the same urge for explanation, and the same naive assumption that the workings of the human mind can be reduced to physical processes.

In this chapter, we will see why this assumption is flawed and look at an alternative explanation of the human mind. We will see that there are some mental phenomena that can't be explained in terms of neurological activity, either because they occur when brain activity is apparently absent, or because they are of an entirely different nature to that which a direct correlation would suggest. We will see that there is a lot of evidence to suggest that it is not the brain that is primary, but the mind, and that psychology cannot be reduced to neurology.

Psychology should not be seen as the study of brain functioning, but of non-material mental phenomena. For example, the best way to examine memory is not in terms of the brain, but in terms of mental functions and structures. The hippocampus doesn't store memories, but there certainly *is* a part of the mind that performs this role. We can't pin down exactly where it is, or what it is, because it doesn't exist in physical space. It exists in mental space, which is just as real as the physical, although it isn't made up solid particles and molecules.

Problems with identifying correlates

If the brain is the source of all mental activity then we would expect a very direct and reliable correspondence between mental processes and neural processes. We would expect to be able to pinpoint certain neural states that correspond exactly (and repeatedly, in every person) with certain states of mind. We would expect to be able to pinpoint precisely the neuronal correlates of states such as love, happiness, depression, aesthetic appreciation, spiritual experiences and so on. We would expect to find precise neuronal correlates for certain types of behaviour and attitude, such as materialism, hedonism, religiosity, optimism, anxiety and so on.

When neuroscientists began to examine the brain, this is what they expected to find (in the same way that some expected to find the neuronal correlates of consciousness itself). And we have found, of course, that certain parts of the brain are generally associated with certain psychological processes. For example, we know that the cortical areas of the brain are associated with learning new activities, and that the left prefrontal cortex is generally associated with planning and controlling our thoughts and actions. We know that the parietal lobe is associated with speech and spatial perception, that the temporal lobe is associated with understanding language, and so on.

However, even after decades of intensive research, neuroscience is nowhere near a precise one-to-one mapping of conscious experience and neural activity. The brain activity associated with certain states of mind is not generalizable from person to person, and it may even vary in the same person, in different circumstances. When a person experiences the feeling of being in love, their brain activity may be different to that of another person who is in love – and even different to the same person's brain activity on another occasion, when they were in love with someone else. Another issue is that animals appear to have some of the same experiences as human beings – pain and fear, for example – which would imply that, at these moments, they experience exactly the same brain states as we do. The brain state of a distressed pig that is being taken off to an abattoir and knows it's going to be killed should be essentially the same as the brain state of a human being who is being robbed at gunpoint, or is told he or she is going to die of cancer. But our brains are so different from those of animals that this seems absurd. According to this reasoning, the experience of pain or fear cannot be identified with any specific brain state.

One of the assumptions of the idea that mental states are the same as brain states is that mental experiences are produced by the firing of neurons in certain parts of the brain. But the correlation between mental states and the firing of neurons is inconsistent. This applies to consciousness itself – as Giulio Tononi has written, the "firing of the same cortical neurons may correlate with consciousness at certain times, but not at others".[4] (Cortical neurons are simply the brain cells in the cerebral cortex, the largest part of the brain.) Neuroscientists have also found that there may be a strange mismatch between the intensity and complexity of a mental state and the amount of neurons or neuronal firings involved. There isn't necessarily any proportionality between the two, and sometimes intense experiences may arise without a significant degree of neuronal firing.[5]

The problem of mental illness

If the brain is the source of all mental experience, we would also expect there to be a clear relationship between mental illnesses and brain states. For example, if a person feels depressed, we would expect there to be precise measurable parameters of neural activity that correspond precisely to the state of depression. We would also presume that it would be possible to "fix" mental illnesses like depression by altering brain activity.

In fact, the latter assumption is one of the most pernicious results of neuromania. Because neuromania sees mental states as brain states, it treats psychological conditions as if they are brain conditions, and assumes that these can be "fixed" with psychoactive drugs. This is the "medical model" of mental illness, which uses drugs to treat conditions like depression, attention deficit disorder or schizophrenia. The model assumes that these conditions are produced by neurological abnormalities, such as the lack or excess of certain brain chemicals, or too little or too much activity in different parts of the brain.

From the materialist point of view, this makes complete sense. If the mind is fundamentally physical in nature, as a product of the brain, then it should be treated in the same way as the rest of the physical body. Psychological problems are brain problems, and brain problems are physical problems, which can be treated by medical interventions. The treatment of psychological issues with psychoactive drugs is therefore, you might say, a direct application of materialism. As the professor of brain science Eric Kandel has put it, "All mental processes are brain processes, and therefore all disorders of mental functioning are biological diseases... Where else could [mental illness] be if not in the brain?"[6]

However, there is, in fact, little agreement about which parts of the brain are associated with mental illnesses. In the case of depression, the best that neuroscientists can say is that, "like other abnormalities of higher mental functions [depression]... seems to be distributed across several brain regions."[7] In other words, the neural correlates of depression are – like those of consciousness itself – widely distributed and difficult to identify.

There is a popular belief that depression is associated with lower levels of serotonin in the brain, but this actually has very little foundation. Writing in the *British Medical Journal* in 2015, the psychiatrist David Healy described how the myth of a connection between depression and serotonin was propagated during the 1990s by drug companies and their marketing representatives, not long after tranquillizers started to be abandoned due to concerns about their addictiveness. In fact, as Healy states, earlier research in the 1960s had already rejected a connection between depression and serotonin, and shown that the anti-depressants known as "selective serotonin-reuptake inhibitors" (SSRIs) were ineffective against the condition. However, propelled by the marketing millions of the pharmaceutical industry, the myth of depression as a "chemical imbalance" that could be restored by medication quickly caught on. It was appealing because of its simplistic portrayal of depression as a medical condition that could be "fixed" in the same way as a physical injury or illness.[8]

Around one in ten Americans take anti-depressants allegedly to "correct" a chemical imbalance in their brains, and increase their serotonin levels. But, not surprisingly, since the link between depression and serotonin is very dubious, the evidence for the efficacy of SSRIs is itself very dubious. Some clinical trails have suggested that anti-depressants can be effective in more serious cases of depression, but they are most often prescribed for mild depression, where they are mostly ineffective and often have serious side effects.[9] Studies led by Irving Kirsch, a professor of medicine at Harvard, have found that anti-depressants bring about only a small, clinically meaningless, improvement in mood compared to placebos. As a result of this research, Kirsch came to the conclusion that depression has nothing to do with a chemical imbalance in the brain, and that anti-depressants such as SSRIs are themselves actually placebos.[10]

Of course, the fact that depression is not related to low levels of serotonin – and can't be fixed by serotonin enhancers – doesn't in itself disprove that depression is caused by brain states. It could be that depression is caused by other types of brain activity, which haven't yet been identified, and it may be that different medication will be developed to counter these. However, it seems much more likely that the reason why it is difficult to precisely identify any neurological patterns associated with depression – and the reason why it is difficult to treat depression with medical interventions – is because the condition does not have its source in the brain. It is an example of a mental state that cannot be reduced to a brain state, and another illustration of the indeterminate relationship between mind and brain. Depression is not the result of a malfunctioning brain, but of environmental, existential and spiritual factors. People don't get depressed because of chemical imbalances in the brain, but because of jobs that are exhausting and unfulfilling, because of a lack of meaning or purpose or love, or a lack of contact with nature, a lack of exercise, or because of too much stress and difficulty, and so on. As a 2017 United Nations statement about the treatment of depression put it, the present-day "biomedical narrative of depression" is based on the "biased and selective use of research outcomes" and "on a reductive neurobiological paradigm". It concluded that the "excessive use of medications and other biomedical interventions… causes more harm than good, undermines the right to health, and must be abandoned".[11]

I'm not saying that there is no relationship between brain activity and depression, and other mental disorders. It may well be that people with depression are more likely to have certain patterns of neurological activity (although these haven't been clearly identified yet). But there are two essential points:

the first is that these patterns of neurological activity are unlikely to be reliable and consistent, from person to person, or even for the same person at different times; the second point is that these patterns of neurological activity – if indeed they exist – should not be seen as the cause of depression but the neurological *register* of depression. In other words, depression should be seen as primarily a mental state caused by a wide range of environmental or existential factors, which then cause changes in brain activity. If this is the case, we would not expect there to be reliable correspondence between brain activity and depression. If the brain does not produce depression, but only registers it, we would expect there to be a good deal of variation.

The adaptable brain

The phenomenon of neuroplasticity also highlights the indeterminate nature of the relationship between the mind and brain.

Neuroscientists used to believe that once an adult human being stops growing their brain remains in the same static state throughout their life, at least until deterioration in old age. In a similar way, scientists used to believe that after a stroke or brain injury a person's brain would remain permanently damaged, and be unable to repair itself. However, it has now become clear that in reality the brain is plastic, and has a great capacity for change and recovery. It is not "hard-wired". We can form new connections in different parts of the brain, and even generate entirely new brain cells. (We will look at some examples of this shortly.)

One of the most interesting aspects of neuroplasticity is that functions can sometimes shift to different parts of the brain. This often happens during recovery from a stroke or other brain injuries. If the part of the brain that used to be associated with a particular activity is damaged, a new uninjured part of the brain will now become linked to it. After a stroke, the brain often reorganizes itself. New connections and pathways to the undamaged parts of the brain are created, so that healthy cells can take over the roles performed previously by cells that have been destroyed. This is another reason why it is problematic to link a particular part of the brain to a particular mental function (or to consciousness itself). As we saw earlier, certain functions may be generally associated with specific parts of the brain, but these associations may shift around after a stroke or brain injury.

This flexibility seems to suggest that the mind utilizes the brain in a similar way to how a musician uses an instrument. For example, imagine a guitarist who is playing a concert when one of his strings breaks in the middle of a song (which happened to me many times when I played music). If they are playing individual notes, they will simply transfer the notes to a different string, played on higher or lower frets. If they're playing chords, they'll probably play the same chords in a different shape, higher or lower up the fretboard, in a way that makes the missing note less noticeable. But they will find some way of adjusting so that the music continues.

Studies of the brain activity of blind people have shown similar findings. There is a part of the brain – the visual cortex in the occipital region – that in most people is associated with visual processing. You might assume that if a person becomes blind, the visual cortex would simply become inactive. But as you can probably guess, this isn't the case. Research has shown that, both for people who were born blind and those who lose their sight later, the visual cortex remains active. It is "co-opted" to help out with other functions.[12]

Even more strikingly, the mind is so adaptable that it can maintain normal levels of mental activity even when large parts of the brain are missing. In 2007 the medical journal *The Lancet* reported the case of a 44-year-old French man who had managed to live a completely normal life despite the fact that there was a huge hole in his skull, where most of the brain normally lies. This was discovered when, for the first time in his life, the man had a brain scan, which showed that his skull was filled with a huge fluid-filled space. His brain tissue was concentrated in a thin sheet around the edges of his skull. The man was married with two children and worked as a civil servant, and until that point there was never any indication of anything abnormal about him. The researchers could identify normal brain structures, such as the frontal, parietal, temporal and occipital lobes, but they were all massively reduced in size.[13]

In a similar case, in 2014, a 24-year-old Chinese woman had a brain scan after complaining of nausea and dizziness. The CAT scan showed that the entire cerebellum was missing from her brain. Where it normally is there was just a space filled with cerebrospinal fluid. The cerebellum is usually seen as the most important part of the brain, containing about half its neurons, so it is difficult to imagine a person being able to function without it. But although the woman had some minor ailments – for example, slurred speech and an unsteady walk – she had otherwise lived a completely normal life. She was married with a daughter and had never suspected that there was anything seriously wrong with her.[14]

Although these cases are rare, many similar ones have been found. In some cases, people have actually been found to be more intelligent than average, despite having large parts of the brain missing. For example, one man had had the entire left hemisphere of his brain removed at the age of five (to control his epileptic seizures) and had grown up to show higher-than-normal levels of intelligence and language skills.[15] Other people have been found to function normally with as little as 5 per cent of the normal amount of brain tissue.[16]

If one takes the conventional view that the mind is produced by the brain, this amazing flexibility is difficult to explain. How could mental functioning remain essentially the same in spite of such massive structural differences? Imagine a computer working in exactly the same way with only 5 per cent of its normal circuitry, or with essential parts of its hardware missing – or a computer that is damaged and can spontaneously shift functions to undamaged parts.

These cases certainly make more sense if we take the spiritual view that the mind comes through the brain rather than from it. While the brain is obviously strongly associated with mental activity, and plays an essential role in its facilitation, we shouldn't expect there to be a direct causal link or a consistent correspondence between brain states and mental states – for the simple reason that the brain does not directly produce mental activity. To return to the metaphor I used earlier in this chapter, the music of the mind comes through the brain, not from it.

CHAPTER 5

HOW THE MIND CAN CHANGE THE BRAIN AND THE BODY: MORE PUZZLES OF THE MIND AND BRAIN

The Victorian scientist TH Huxley – the grandfather of the author Aldous Huxley, whose spiritual views he would surely have abhorred – was the Richard Dawkins of his time. He espoused his vision of a mindless, machine-like world with the passion of a religious zealot. He defended Darwin's theory of evolution so fervently that he was known as "Darwin's bulldog". (It was actually Huxley, not Darwin, who coined the phrase the "survival of the fittest".)

Huxley was convinced that consciousness is just a by-product of the brain, comparing it to a steam whistle produced by a locomotive engine. This led him to conclude that human beings' sense of volition is an illusion, and that consciousness was "completely without power of modifying" the functioning of the body in the same way that the steam whistle could not alter the functioning of the engine. As Huxley continued, "Have we any reason to believe that a feeling, or state of consciousness, is capable of directly affecting the motion of even the smallest conceivable molecule of matter?"[1]

80

From the materialist point of view, this makes complete sense. If the mind is just a product of matter, how could it change or influence matter? How could it have any effect on the molecules of the body and the brain? To use a modern metaphor, how could the images on the screen of a computer change the software of the computer?

But Huxley was wrong. In actual fact, the mind can alter the matter of the brain and the body in profound ways, casting further doubt on the materialist assumption that the mind is produced by the brain.

How the mind can change the brain

This is another reason why neuroplasticity is so important. Often – as we saw in the last chapter – it occurs spontaneously, in response to brain injuries. But it also often occurs in response to mental activity, or conscious mental training.

If you ever decide to become a taxi driver in London, it's more difficult than in comparable capital cities such as New York or Berlin. Even after the advent of satellite navigation, prospective London cabbies still have to pass an exam called the "Knowledge" that requires them to learn hundreds of routes across the city, involving 25,000 streets and 20,000 landmarks. The process normally takes at least two years, and only around half of those who sit the exam pass.

We have seen that memory is associated with the hippocampus, even if memories are not actually stored there (and even if it involves other brain areas too). The neuroscientist Eleanor Maguire has spent several years studying the brains of London taxi drivers, and concluded that they have significantly more grey matter in their posterior hippocampi than normal. At first, her studies didn't make it clear whether this had come about as a result of their training, or whether the selection process simply favoured candidates with larger than normal hippocampi. However, further research tracked how the drivers' brains developed over time and showed clearly that the increase in grey matter was a result of their training. This is probably due to both an increasing number of connections between existing cells and the genesis of entirely new cells.[2]

There have been similar findings with groups of medical students, whose brains were shown to change significantly through weeks of intensive study before their exams.[3] Studies of the effects of meditation have shown that significant neurological changes can occur within a period of only eight weeks. In a 2011 study, a group of 16 new meditators were given brain scans before and after an eight-week meditation programme. After meditating for an average of 27 minutes each day for those eight weeks, the scans showed a significantly greater amount of grey matter throughout many regions of the brain.[4] Similarly, a 2017 study found that cognitive behavioural therapy (CBT) can cause significant neurological changes. Focusing on a group of people who were suffering from psychosis, researchers found that a 12-week course of CBT brought about increased connections in the amygdala and frontal lobes, which the researchers associated with rational thinking.[5]

How could any of this be possible if mental phenomena are just the product of brain activity? How can an epiphenomenon change the nature of the primary thing it's just a by-product of? In materialism, the mind is seen as a shadow of the brain. And how can a shadow change the object that it's the shadow of? Or in Huxley's analogy, how could a steam whistle change the nature of the engine itself?

One could argue that parts of the brain have simply exercised themselves more, and as a result have grown like a muscle. But this is not an appropriate analogy. No one knows how the brain generates new cells and forms new connections between cells, but it certainly isn't in any way similar to muscle growth, which occurs through the damage and repair of muscle fibres. Muscle growth certainly doesn't involve any increasing interconnectivity between cells either. And in any case, even if we accept the exercise analogy, it still doesn't change – or explain – the fact that mental intention and activity are bringing about major structural changes to the brain, in a way that suggests the primacy of the mind.

I'm not saying that the brain does not affect the mind. This would be absurd. As I have mentioned in relation to consciousness, it is clear that when brain functioning is changed by drugs there are significant changes to mental experience. And the same is true of injuries to the brain, which can profoundly affect – and impair – a person's mental functioning. What I am suggesting is that there is a two-way relationship between the mind and brain, and that the mind can affect the brain in profound ways, just as the brain can affect the mind. And as I mentioned in the last chapter in relation to depression, although there may be associations between neurological functioning and certain mental states (as the earlier examples of neuroplasticity show), we should expect there to be a great deal of variation in these. It may be that this depends on the direction of causation. When the primary change happens at the level of the brain – for example, when brain activity is altered through drugs – then it may be possible to identify precisely a correlation between neurological activity and mental experience. But when the primary change happens at the level of the mind – for example, when a person gets depressed because a relationship ends or because they're stuck in a frustrating job – then it is more likely that there is just a general association, without much reliability or consistency.

How the mind can change the body

The primacy of the mind is suggested in an even more startling way by its amazing capacity to change the form and functioning of the body.

The most common example of this is the placebo effect. This phenomenon was first noted in the mid-20th century when scientists began to use double-blind randomized controlled trials in research. This meant that a procedure or treatment would be carried out with two groups: one group that would actually receive the treatment, and another – the "control group" – that would receive inert substances (with neither group knowing which they had received). In medical trials, researchers began to find that, for some unknown reason, the control groups would report some effect: that their symptoms had alleviated or that pain had faded away, even though they had actually had no treatment. In some cases the control group would even report negative side effects of the treatment (these became known as nocebo effects). Over the course of decades of research, a vast array of different conditions have been shown to benefit from placebos, including acne, Crohn's disease, epilepsy, erectile dysfunction, ulcers, multiple sclerosis, osteoarthritis, rheumatism and colitis.

At first, scientists assumed that the placebo effect was just a subjective phenomenon. They believed that people were just imagining that their symptoms had improved, without any real physiological changes taking place. (Even now, some sceptics suggest this.) But it became apparent that this wasn't the case. Even though placebo patients were receiving inert substances with no physiological effects, real and measurable physiological changes were occurring. This was evident in a 1978 study in which 40 patients were given a placebo painkiller following dental treatment. Shortly afterwards, they were divided into two groups, one of which received another placebo painkiller, while the other received naloxene, a substance that stops the release of pain-numbing endorphins in the brain. The second group reported significantly more pain than the first group, suggesting that the original placebo produced chemical changes, and that these were being blocked by the dose of naloxene. This seemed to show that placebos could bring about the same chemical changes as actual drugs.[6]

Later, brain-imaging technology verified the physiological effects of placebos. Studies found that, when taken as painkillers, placebos caused decreased activity in parts of the brain associated with pain. They also made use of many of the same neurotransmitters and neural pathways as opioids and marijuana. In other words, the placebos were actually causing the release of endogenous chemicals in the brain. In a similar way, researchers found that placebos could cause the release of dopamine in the brain when taken by patients with Parkinson's disease. Other studies have shown similar neurological changes relating to depression, anxiety and fatigue, and other physiological changes such as blood pressure, heart rate and the activation of the immune system.[7] (As we saw in the last chapter, it is also possible that anti-depressants largely function as placebos.)

Interestingly, research has also shown that for a placebo to work it may not even be necessary to deceive patients into thinking they are taking an active substance: placebos can still work even when we know that we're taking them. In a study published in 2010, Ted Kaptchuk, a professor of medicine at Harvard, compared two groups of patients suffering from irritable bowel syndrome. While one group didn't receive any treatment, the other was given placebos. Unusually, members of the second group were told that they were taking fake drugs; to emphasize the point, the bottles of pills were labelled "placebo pills". Kaptchuk was surprised to find that, despite this, the members of the second group reported significant relief of symptoms. This seems bizarre, although it is probably significant that Kaptchuk's team was careful to tell the patients that placebos often have healing effects.

Sham surgery

In 2004 a radiologist at the Mayo Clinic in Minnesota – one of the United States' most prestigious hospitals – named Dr David Kallmes decided to try an unusual experiment. For many years, he had been performing an operation called a vertebroplasty, in which broken backs are healed through the injection of a medical cement. The procedure had always been very successful, relieving severe pain and allowing people to walk and exercise without difficulty. However, one thing had always puzzled Dr Kallmes: occasionally the operation would go wrong (for example, if cement was injected into the wrong vertebra), but patents would still appear to get better.

In order to investigate this further, Kallmes conducted a trial of 131 patients, in which half received a real vertebroplasty, while the others had a fake operation. In the latter, patients were wheeled into the operating theatre and given an anaesthetic, but rather than being injected with the cement, they were simply pressed hard on the back. Amazingly, the results found that both groups experienced the same amount of pain relief and the same amount of improvement in function – that is, in walking, climbing stairs and other forms of exercise.[8]

This is an example of another bizarre aspect of the placebo effect known as "sham surgery". This is when surgeons literally pretend to do an operation, doing everything they would normally do – for example, making an incision, picking up instruments, giving instructions to colleagues, then closing the incision – but without actually making an intervention. Although this seems to defy common sense, many trials of sham surgery have yielded positive results. In a Finnish study published in 2013, sham surgery was performed on patients suffering from torn knee ligaments, and in severe pain. Even though the sham surgery patients were anaesthetized, surgeons went through the whole ritual of an operation in meticulous detail, passing instruments and making the normal sounds associated with an operation. But again, the incision was closed without any procedure being carried out. Some patients received real treatment too, and the results were compared. Once again, no significant difference was found between the two groups. Patients who had had sham surgery reported the same degree of pain relief and improved function.[9]

Shortly after this trial, another team of researchers published a comprehensive review of every recorded trial of sham surgery, and found 53 cases where it was practised alongside normal surgical procedures. They found that sham surgery was beneficial in 74 per cent of trials – in half of these, to the same degree as the actual procedure. In some cases, it was found to be more beneficial than the actual procedure.[10]

Not surprisingly, one conclusion that has been drawn from the success of sham surgery is that a lot of unnecessary operations are being carried out. And indeed, since the publication of these findings, US insurance companies have been less willing to fund operations such as vertebroplasties. However, the findings obviously don't imply that real vertebroplasties do not work, or are not necessary – only that the healing powers of the subconscious mind can simulate the effects of the real procedure.

The placebo effect in general has become so familiar nowadays that we may need to remind ourselves of how bizarre it really is. Isn't it incredibly strange that healing and pain relief can take place without any actual treatment? It appears that, even now, most scientists don't grasp the full implications of the placebo effect: that the human mind has the capacity to powerfully influence almost any aspect of our physiology, including the healing of many conditions and the alleviation of a large range of symptoms. And, in turn, this implies that the mind is actually primary to the physical stuff of our bodies and brains, rather than a by-product of brain activity.

Perhaps this is why many scientists don't contemplate the implications of the placebo effect – because those implications are so problematic for the conventional materialist model that many of them have adopted.

Hypnotic healing

One of the strange things about the placebo effect is that it doesn't work consciously. We can't simply will ourselves to change. We can't simply think: "I wish I wasn't ill – I think I'll try to make myself better." The power of the mind to change the body can only be activated at a subconscious level, through the belief that a change is taking place (in most cases, pain relief or healing). Belief seems to uncover a strange capacity of the mind to alter the physical structures and processes of the body.

These powers can also be activated in a state of hypnosis. The suggestions of a hypnotist can connect to the subconscious mind in the same way that our own beliefs can.

The hypnotic state is still mysterious – there is no clear explanation of what happens when a person becomes hypnotized, or how the state is different from normal consciousness. But the essential aspect seems to be that, under hypnosis, the normal conscious ego becomes immobilized. Ego functions such as volition and control are taken over by the hypnotist. And with the ego in abeyance, the hypnotist has direct access to the person's subconscious mind. This is the principle of hypnotherapy – the hypnotist can work directly with the subconscious, addressing the fears and phobias that are rooted there. And by the same means the suggestions of a hypnotist can bring about changes not only to the form and functioning of the brain but also the body as a whole.

Hypnosis is especially effective as an analgesic. In the 1840s a Scottish doctor living in India named James Esdaile began to use hypnotism for pain relief in operations. In his region of India, many men would develop enormous tumours (weighing up to 100 pounds/45 kilograms) in the scrotum – due to infection from mosquito bites – and the operation to remove them was so painful that men would often put it off for years, only having it as a last resort. Esdaile began to use hypnotism (or "mesmerism", as he referred to it) as a way of relaxing patients, so that they would agree to have the operation. And to his surprise, he found that they didn't feel any pain during the operation. He also reported that, in some cases, there was no pain or injury after the operation either, and that the healing process was faster. As he wrote, "less constitutional disturbance has followed than under ordinary circumstances. There has not been a death among the cases operated on."[11]

Word began to spread about this amazing surgeon who could remove the massive tumours in 20 minutes without pain or after-effects, and soon patients began to flock to Esdaile's hospital near Calcutta. Esdaile began to use hypnotism in other procedures too, including eye surgery, the removal of tonsils, breast tumours and childbirth. Esdaile was sure that it wasn't just a matter of his patients pretending (to themselves and/or to him) that they weren't feeling any pain – he noted that, in addition to a lack of writhing and moaning, patients didn't display physiological signs of pain such as changes to their pulse rate and eye pupils. In the mid-19th century mortality rates for operations were incredibly high: a staggering 50 per cent of patients died during or after them. But in 161 recorded cases of Esdaile's operations, the mortality rate was only 5 per cent. The reasons for this are unclear. Esdaile himself believed it was due to "vital mesmeric fluids" passing from him to the patient, which stimulated the healing process. However, it was probably related to reduced loss of blood, and perhaps an activation of the same self-healing abilities that can occur with a placebo.

There are many cases of hypnosis being used by physicians before and after Esdaile, but the practice was soon superseded by the use of chemical anaesthetics. However, there were still some areas of medicine where the practice continued – dentistry, in particular. At the turn of the 20th century, hypnosis was dentists' main method of pain management, and the practice became almost universal for dentists during the two world wars, when chemical anaesthetics were scarce and facial trauma was common. Even now, some dentists still use hypnosis, especially in cases where a person's medical history precludes the use of an anaesthetic. Recent research with patients who had teeth extracted under hypnosis showed that "hypnotic-focused analgesia" can increase pain thresholds by up to 220 per cent. This research also found that 93 per cent of patients experienced reduced postoperative pain and haemorrhage.[12] (This obviously links to Esdaile's finding that mortality rates decreased very sharply as a result of his use of hypnosis. Clearly, hypnosis reduces blood loss and haemorrhage.)

As with the placebo effect, research has shown that the analgesic effect of hypnosis isn't simply a subjective phenomenon but something that brings real physiological changes associated with an absence of pain – for example, slow rates of pulse and respiration, and changes to galvanic skin response. And again, as with the placebo effect, brain-imaging has indicated decreased activity in the brain areas associated with pain. This might suggest that hypnosis can activate the same pain-numbing endogenous brain chemicals as the placebo effect. However, this doesn't appear to be the case. Earlier in this chapter an experiment was described that showed the pain-numbing effects of a placebo were negated by naloxene, which blocks the release of endogenous opioids. However, a similar experiment found that, under hypnosis, the naloxene had no effect, which suggests that endogenous opioids weren't involved. As with so many aspects of hypnosis, its pain-numbing capacity remains unexplained.[13]

Hypnotism has been shown to be effective in many other areas besides pain management. In the first half of the 19th century it was used by physicians as a treatment for illness, and found to be effective against conditions such as epilepsy, neuralgia and rheumatism. However, the area where it has been found to be most effective is skin conditions. In highly suggestible people, hypnosis has been used to rapidly heal wounds and burns, to make warts and blisters disappear and to control the bleeding of haemophiliacs. Conversely, highly suggestible people may produce blisters or burn marks if they are told by a hypnotist that their skin has been burned, or if they are blindfolded and the hypnotist pretends to touch them with a red hot poker or another object.[14]

Hypnosis can also bring about significant structural changes to the body. In the 1970s a number of studies (involving a total of 70 women) found that hypnotic suggestion could enable women to increase their breast size. While under hypnotic trance, the women were given a variety of different suggestions and visualizations. For example, they were asked to visualize how they wanted to look, or to feel warm, tingling sensations in their breasts. After 12 weeks, 85 per cent reported an increase in breast size, by an average of 1½–2 inches. Follow-up studies showed that the size increase was permanent for 81 per cent of those who developed larger breasts.[15]

Even though the effects of hypnosis are unexplained, it seems clear that the process involves the same innate, self-healing and self-regulating powers that are harnessed through the placebo effect. The hypnotist really does nothing other than create the conditions under which these powers can be activated.

Other puzzles of the mind influencing the body

This discussion of human beings' incredible self-healing abilities may throw some light on controversial practices such as faith healing and homeopathy. From a materialist perspective these practices make no sense at all. Materialist scientists often express incredulity that homeopaths believe that inert solutions (diluted so heavily that no molecules of an active substance remain) can bring about healing. But this is exactly what happens with the placebo effect, which most materialists accept is real. So perhaps these practices are the result of the placebo effect? If a person believes that a therapy or technique can heal them, then there is a chance that they will be healed through their own self-healing abilities. In suggesting this, I don't mean to belittle these healing practices. As we have seen, the placebo effect can be enormously effective. It may be that such practices are simply different ways of harnessing the same self-healing capacity as the placebo effect. And in this sense, they may be extremely valuable. (It also possible that homeopathy and other healing practices are more than placebos, and that they operate via forces or phenomena that we are unaware of – and perhaps will never discover.)

Another possibility – particularly in relation to faith healing, and also to "distance healing" (for example, when a number of people remotely pray for the recovery of a sick person) – is that if we as individuals are capable of influencing the form and functioning of our own bodies, perhaps we are able to influence the form and functioning of other people's bodies too? My feeling is that this is potentially possible, but probably much less common. There's a massive difference between being able to influence our own physiological functioning – which is intimately bound up with our own mental activity, and can't be separated from it – to being able to influence another person's physiological functioning, which is much further removed from us.

There may be a connection here with the psi phenomenon of psychokinesis, which is when a person shows the ability to influence the form and functioning of material objects through mental influence. (For example, when a person manages to stop a clock ticking, or bend the hands of a watch without physical intervention.) If mind is primary to matter, and pervades it, then it is potentially possible that mental influence can influence matter. But again, this is probably much less common than causing changes in one's own body – and probably less common than bringing about changes in other people's bodies too due to the lack of intimate connection between material forms beyond our body.

There are so many other examples of how beliefs can affect the form and functioning of the body that one could easily fill a whole book with them. And since I think the evidence I've presented is convincing enough already, I don't want to bore the reader with a bombardment of unnecessary material. But before we end this chapter, let me briefly mention two (or three, if you regard phantom pregnancy separately) other phenomena that I believe add significant weight to the case I'm making.

First of all, it's important to look at the opposite of healing: making ourselves ill. It stands to reason that if subconscious beliefs and suggestions can alleviate pain and induce healing, they should also be able to create (or increase) pain and induce illness. If we can heal ourselves, we can surely also make ourselves sick? The technical term for this is psychogenesis: the generation or exacerbation of illness through mental factors. There is obviously a close connection with the concept of "psychosomatic" illness, although doctors tend to restrict this term to psychological factors influencing diseases rather than actually generating them.

Nocebos are the most obvious example of psychogenesis. Research from random-controlled trials with medication for Parkinson's disease showed that almost two-thirds of patients experienced negative side effects from a placebo. For one-tenth of patients, these were so severe that they couldn't continue with the trial.[16] In another study there was a striking example of the nocebo and the placebo effect working together when asthmatic patients were asked to inhale an inert substance (a saline solution) that they were told was an allergen. This led to a significant increase in airway resistance, which brought about asthma attacks in some patients. Fortunately, the attacks receded when they were given the saline solution again – only this time they had been told it was an anti-allergen.[17] A final dramatic example of the nocebo effect is provided by the case of a young American man – whose girlfriend had just ended their relationship – who decided to commit suicide at a time when he was taking part in a clinical drugs trial. He took 26 capsules of what he believed were anti-depressants, although in fact these were placebos from the trial. He was rushed to hospital, where medics found that his blood pressure had dropped to a dangerously low level. As an emergency measure he was given intravenous fluid to restore his blood pressure to a normal level. But then medics realized that he had only taken placebo tablets, and soon his blood pressure returned to normal.[18]

In a more general sense, it's very likely that a significant proportion of illnesses we suffer from are psychogenic or psychosomatic. Studies from around the world have suggested that between 25 per cent and 50 per cent of patients who visit primary care doctors have "medically unexplained symptoms" caused by psychological factors.[19] And it's interesting that skin conditions – such as psoriasis and eczema – are known to have a particularly strong psychosomatic element. This fits with the finding mentioned earlier that hypnotic healing works especially well with skin conditions. (Other conditions that are strongly psychosomatic are stomach ulcers, heart conditions and high blood pressure.)

One of the most significant psychogenic phenomena is false (or phantom) pregnancy. Nowadays cases of false pregnancy are quite rare, due to pregnancy tests, increased general medical knowledge and less social pressure for women to become pregnant. But in the 1940s it was estimated that one in 250 pregnancies was false. As with the placebo effect, false pregnancy isn't just a matter of a woman convincing herself that she's pregnant; it involves many of the same physiological changes as actual pregnancy, including a cessation of periods for months, an enlarged stomach (that grows at roughly the same rate as a normal pregnancy), sore breasts that secrete colostrum (or pre-milk) and nausea. In some cases, actual pregnancy hormones are released, such as oestrogen and prolactin. There may also be fetal movements, experienced not only by the mother but also by others, including doctors.[20]

Before pregnancy tests existed these physiological changes were often so similar to actual pregnancy that doctors couldn't tell them apart. A 1951 study of false pregnancy, as reported in the *Journal of the American Medical Association*, stated that the symptoms were so convincing that they "may tax the diagnosis abilities of the ablest physicians".[21] A study of 27 instances of false pregnancy from 1937 to 1952 found that, in every case, at least one doctor agreed with the woman's conviction that she was pregnant.[22]

To end this survey of how the mind can influence the body, let me draw your attention to one last strange phenomenon: the ability to postpone death. If we are able to change the form of our bodies, and to heal or generate illness, then is it possible that we have the ultimate form of control over our bodies by being able to choose when their functioning ceases? This does seem to be the case, within certain limits. Obviously, we can't delay death indefinitely. But if we are in the process of dying, we appear to have some control over when death occurs. Usually this occurs in relation to occasions like birthdays, anniversaries, cultural occasions (like Christmas or Eid) or the arrival of a relative or friend. I advertised for some examples of this on social media and received a large number of reports. One person described how her mother had stage-four cancer and hadn't eaten (even via intravenous feeding) for over a month. The doctors were amazed that she was alive, but she told them she was waiting to see her son from America (she lived in Europe). Once her son arrived, she told the doctors, "I will let go as soon as he leaves." She died a few hours after his visit. Other people told me stories of people waiting for occasions such as the birth of a grandchild, a marriage or graduation. An acquaintance of mine who worked as a chaplain in a hospice told me that he frequently witnessed people waiting for the final visits of best friends, lovers, relatives – or for the death of someone else who was ill (in one case, even the death of a dog).

There is empirical evidence of this phenomenon too. Studies have found that rates of mortality decrease prior to culturally significant events (like Christmas or Passover) and then increase afterwards. One 1990 study found that mortality among Chinese people fell by 35 per cent during the week before the Harvest Moon Festival and peaked by the same amount during the following week.[23] This suggests that a strong intention can have an influence over the biological process of death.

A variation of this phenomenon is often reported by nurses and other caregivers. If a person is close to death, and relatives gather at their bedside, it is not uncommon for them to delay their moment of passing until the relatives leave the room. As one nurse told me, "I've seen many a person hang on until a beloved relative could come and say goodbye or a special time of year had passed. But the opposite also happened... Sometimes the person would slip away quietly, when their caregiver/relative wasn't present." Another ex-nurse told me: "The most common examples that I have witnessed have been the dying person deciding to die when relatives and friends weren't about." It's not clear why people should choose to pass away when their relatives are out of the room. Perhaps they want to save their relatives or caregivers the emotional pain of watching them die, or to avoid the drama of their emotional reactions.

Implications

All of the examples we've looked at in this chapter make it clear that TH Huxley was completely wrong when he stated that the mind does not have the power to "modify" the body. Since it is beyond doubt that this happens, what we really need to consider are its implications.

One implication is that we have much more control of our own health, and our own bodies, than we normally assume. We have a massive potential for self-regulation and self-healing that we are barely aware of. Of course, this has implications for healthcare professionals too. At present – even in the face of massive evidence of the influence of the mind provided by the placebo effect – the medical profession is dominated by a materialist conception of the body as a machine. But, clearly, the influence of beliefs and intention should be taken account of, especially when these can be harnessed to such a positive effect.

This also implies the importance of context, and of the relationship between practitioner and patient. If the body is just a machine these factors can be disregarded as incidental. It doesn't matter how a mechanic goes about fixing your car, or whether you have a good relationship with him, as long as the car gets fixed. But since a patient's mental state has so much influence over the health of their body, it is important for healthcare professionals to be attentive to their patients' psychological well-being, and to nurture a positive relationship with them. Best of all, they should make a conscious effort to harness every patient's innate self-healing abilities.

In a wider sense, in terms of the argument of this book, it is important to consider that all of the phenomena we've looked at over the last two chapters – such as the placebo effect, hypnotic healing, psychogenesis, phantom pregnancy and the ability to postpone death – *cannot be explained from a materialist perspective*. For a materialist, these are simply rogue phenomena which make no sense, and have to be disregarded or explained away. But because their existence cannot be doubted, we have to conclude that it is the materialist model itself that makes no sense and should be disregarded.

However, from a spiritual perspective these phenomena *can* be explained. They do make sense if we presume that mind is more fundamental than the matter of the body, and so has the capacity to alter its form and functioning. From the point of view of panspiritism, human consciousness is essentially an influx of the spiritual energy of the universe. The brain canalizes universal consciousness into our own individual beings. In this way the mind can be seen as a product of consciousness-in-itself interacting with the brain. With its billions of neurons and its incredibly complex interconnections, the brain facilitates mental functions such as memory, information processing, concentration, abstract or logical cognition, and so on. You might say that once it has been canalized, the brain enables consciousness to be organized into various psychological functions and processes. The essence of the mind is therefore not the brain, but spiritual energy (or universal consciousness). The brain is simply the facilitator of mind.

Remember that this isn't a dualistic perspective. I'm not saying that mind and the body – or mind and matter – are distinct. I'm saying that mind and the body are different aspects of consciousness, or spirit. But the mind is a more subtle and fuller expression of spirit than matter. It is, you might say, a higher-order expression of spirit. Matter *has* consciousness – since it is pervaded with consciousness – but it is not conscious in itself. But when matter is arranged in complex and intricate ways – such as in cells and organisms – it can allow the canalization of consciousness, so that organisms become individually conscious. And the more complex and intricate an arrangement becomes, the more intense and expansive its consciousness becomes. Some philosophers have referred to this process as involution – meaning that as evolution progresses, consciousness becomes more *involved* in matter. So as evolution progresses, there is an increasing *spiritualization* of matter: living beings become a fuller expression of spirit and move closer to the source from which they, and all things, came.

But the really important point is that because the mind is a more subtle and fuller expression of consciousness, it is primary to matter and to the physical stuff of the body.

Expectations

We therefore shouldn't be surprised that there are no exact correspondences between brain states and mental states. We should expect brain functioning to be very flexible, because its role is not to produce mental states but to facilitate them. We shouldn't be surprised if different mental functions can switch to different parts of the brain in certain circumstances, or even that the brain can facilitate normal mental functioning in the absence of a large amount of brain tissue. Once the canalization of consciousness has occurred and been organized into mental functions there is a great deal of room for flexibility. There is no strict dependence between certain parts of the brain (or certain neuronal networks) and certain psychological processes or functions, even if some may be commonly associated with them.

In a related way, we shouldn't be surprised if it is difficult to specify the precise location of any particular mental function. In the last two chapters we've seen repeatedly that certain functions are associated with certain parts of the brain, but we should remember that the brain works holistically. In actual fact, every function involves many different parts working together, even if one area may be predominant. Again, this is exactly what we would expect of a brain that facilitates rather than produces mind.

We also shouldn't be surprised if the activities and functions of the mind can change the structure of our brains, and have such a powerful effect on the health and condition of the body. Because it is primary and essential the mind is much more powerful than the materialist model assumes. The physical stuff of our bodies is itself a manifestation of universal consciousness (or spirit) and is pervaded with it, and so is very susceptible to the influence of our individual minds.

In the next chapter we're going to continue this discussion further by looking into another phenomenon that suggests – in an even more dramatic way – that the mind (or consciousness) is not dependent on the brain, and can even continue in the absence of brain activity.

CHAPTER 6

THE PUZZLE OF NEAR-DEATH EXPERIENCES

About 15 years ago I worked with a college tutor called John who had had a heart transplant a few years earlier. He had suffered from a condition called cardiomyothopy, a degeneration of the walls of the heart chambers, which had caused the death of his father at the age of 59. John developed the condition at the age of only 32, and his life was saved by a pioneering transplant unit at a local hospital.

During the transplant operation John was surprised to suddenly find himself awake and alert, looking down on his own body from above. He could see the surgeon and the nurses performing the procedure and sensed from their behaviour that there was an emergency – he could see them rushing around, looking anxious. He was also surprised to hear classical music in the operating theatre. He felt himself floating further away from his body, into a darkness that felt strangely peaceful, towards a bright light in the distance. Then he encountered his deceased father. His father seemed surprised to encounter him, and told him, "You shouldn't be here – it's not your time yet." Then John felt himself moving back down towards his body, and lost consciousness. The next thing he knew, he was awake in recovery.

Soon afterwards, when the surgeon arrived to check on his progress, John said to him, "I liked the classical music you were playing in the operating theatre." The surgeon was amazed that he knew this, because John had been unconscious when they turned the music on, and had remained completely unconscious during the operation.

John didn't know what to make of this experience. He told me that he was a rationalist, who was interested in science and didn't believe in any "new age" nonsense. He had tried to tell a few friends and relatives about it but they thought he was crazy. He told me I was the first person who had taken him seriously.

"That's a near-death experience," I told him. "It happens to a lot of people. People die for a short time – their hearts stop beating, their brains don't have any activity, they don't have any of the vital signs. But they continue to be conscious. They leave their bodies, and sometimes encounter deceased relatives."

I gave John a book on the subject but unfortunately lost touch with him soon after that. Hopefully, the book helped him to make sense of the experience, to accept it and integrate it into his life. (Sadly, I discovered recently that John died several years ago, due to complications from his transplant.)

Max's experience

A student of mine named Max had a similar experience, although it happened while he was in a coma rather than under a general anaesthetic. In his late teens, Max was misdiagnosed with an illness and given inappropriate medication, which led to a seizure. (Doctors later explained it as a chemical reaction to amitriptyline and tramadol.) He went on to write his third-year dissertation on near-death experiences, where he described his own experience in detail, which is so remarkable that I will quote from his account at length:

What first began as intense, sharp pains soaring through my frontal lobe would later be explained as the result of oxygen deprivation during a series of seizures. I felt like my head was being overwhelmed, at which moment I felt that my life was over. I did not regain consciousness until a few days later in a hospital Intensive Care Unit. However, during the period of unconsciousness, whilst in a medically induced coma, I experienced an array of experiences that I now understand to be synonymous with and typical of near-death experiences.

What would occur after my searing headaches was an out-of-body experience, where I watched myself from all corners of the hospital room simultaneously. I observed doctors and nurses regularly checking my vital signs and running a variety of tests on my unresponsive body. I witnessed them adjusting wires attached to what I can only describe as a metal helmet connected to a monitor and secondly to inject my stomach with an unknown substance. I recall an immediate sense of fear and panic driven by awareness that whilst I could observe the scene, I could not act with any agency or communicate with doctors or engage my body to respond. I felt disconnected from my physical body, and while I could see the doctors and nurses speaking, I could not distinguish meaning from their words, which I only heard as muffled. This added to my sense of anxiety, as I could not break down the barriers of language, or equally, make sense of the experience as a whole.

As I was in this projective state I was suddenly over-whelmed, absorbed into and engulfed by a bright light that dominated the scene of the room. It engulfed both my physical and non-physical self. I experienced an immediate shift from my feelings of panic and fear to a state of deep peace. I could not feel my physical self; my body felt as if it was no longer here. It was as if I was no longer present. I felt removed from worry and filled with a sense of absolute tranquillity. My only awareness was of the brightness of the light, filling me with peace and nourishment.

It was at this point that I had an encounter with a familiar face, whom I understood to be my great-grandmother, who died early in my childhood. Her

presence was somehow not surprising. She explained
that this experience was unique, and that as in a lucid
dream I would have control over my physical self and be
able to regain consciousness once I was ready. After what
seemed like years of conversation with her I felt ready
to wake up and before I knew it, I was in the hospital
and encountering a man in my room drawing back the
curtains from around my bed and explaining where I was.
Although disorientated and blurry, I was fully conscious
and in control of my body again. My immediate reaction
was to look to where my great grandmother had been
sitting. My last memory was of her informing me to not
be afraid of the darkness in my eye, as I was blind in my
left eye.

One of the most remarkable things about this experience is
that Max's great-grandmother was Greek and only spoke that
language, so that as a child he had never been able to speak
to her directly. (His father had translated between them.) She
spoke Greek to him during this experience too, but somehow
Max was able to understand her. Most strikingly, his great-
grandmother told him information about his grandparents and
his father that he later verified with his family. She described
her upbringing in Greece, told him how she had met his great-
grandfather and talked about their lives together, including
their suffering during and after the Second World War.

In contrast to John, Max's experience had a powerful
impact on him straight away. As he describes it: "This NDE
caused a life-altering transition that would shift my focus and
attention towards a new set of goals, values and openness,
which I still have to this day… [It] has propelled me towards
a sense of higher purpose and a personal truth; knowledge
and acceptance of life after death." This manifested itself in
major changes to his lifestyle. Before the seizure, he had been
studying economics at university, but no longer felt able to do
this. Instead, he wanted to investigate alternative perspectives
and understand the human mind, which is how he came to

study our university course. Before his NDE he was a fairly typical student who enjoyed socializing and didn't think about much beyond having a good time in the present moment. Now he was very happy to spend time reflecting, loved studying and felt a greater sense of purpose.

What are near-death experiences?

In popular discourse the term "near-death experience" (or NDE) is sometimes used to refer to any close brush with death – perhaps due to an accident, a drug overdose or an illness. But in the strictest use of the term, a near-death experience is when a person appears to be clinically "dead" for a short period – when their heart stops beating, their brain registers no sign of activity and the other "vital signs" indicate death – and yet they subsequently report a continuation of consciousness. This may happen following a cardiac arrest, for example. For a few seconds or minutes a person may show no biological signs of life, and yet when they are resuscitated they report a series of unusual experiences.

Researchers have found that there is a "core" near-death experience. It begins with a feeling of separation from the body (or out-of-body experience), sometimes with a humming or whistling sound. Then there is typically a journey through a dark passage or tunnel towards a place of light. There is a feeling of serenity and intense well-being, a sense of calmness and wholeness, which is often so pleasant that some people are reluctant to return to their bodies, and even feel disappointed when they regain consciousness. Often people meet deceased relatives (as both John and my student did) or beings of light. In a smaller proportion of cases there is a "life-review", in which the significant events of a person's life are replayed. Although it is uncommon for every single one of these characteristics to occur in a near-death experience, most of them usually occur.

Throughout the experience, people feel that their senses have become heightened – everything they experience has a quality of intense realness. In contrast to dreams or hallucinations, it feels much more real than our ordinary experience. There is often also a sense of being outside time. Even though a person may only be unconscious for a few seconds, they may undergo a complex succession of experiences that appear to last for hours. There is also a sense of connectedness, or unity. The sense of being a separate entity, enclosed within our own mental space, is replaced by a sense of being part of an interconnected network of being, of sharing identity with other people, or the world in general.

In a survey of 300 NDEs by one of Britain's leading NDE researchers, the psychologist Peter Fenwick, 88 per cent of people experienced feelings of serenity and joy in NDEs. One heart attack victim, who watched from above while paramedics tried to restart his heart and then passed through a tunnel towards a light, commented: "There is no comparable place in physical reality to experience such total awareness. The love, protection, joy, giving, sharing and being that I experienced in the Light at that moment was absolutely overwhelming and pure in its essence."[1] A woman who had a near-death experience during childbirth said that the most striking aspect of the experience was "the absolute peace, the oneness, the completeness".[2]

The light that people see in NDEs is very different to the type of light we experience in our normal lives. It has an extremely radiant, even translucent quality. It gets brighter and brighter as the person moves towards it, but no matter how bright it becomes it doesn't hurt the eyes. At the same time it has a welcoming and benevolent quality. As one person reported, "It was just pure consciousness. And this enormously bright light seemed to cradle me. I just seemed to exist in it and be part of it and be nurtured by it and the feeling just became more and more ecstatic and glorious and perfect."[3]

It is important to mention that some researchers use a wider definition than others of the term "near-death experience", and include experiences in which a person may not be clinically dead. Coma experiences such as Max's are an example of this. In a coma, a person may not have biological signs of death (or even be in immediate danger of dying) although they could be considered close to death in the sense that they are deeply unconscious, with much-reduced neurological activity and severely impaired biological functioning. But as with Max, they may still experience some of the core features of the near-death experience, with very intense conscious experiences that are incompatible with their very low level of brain activity. (Eben Alexander's book *Proof of Heaven* describes this type of NDE.)

This wider definition also includes accidents such as falls, when a person may experience some of the core characteristics of NDEs during the few seconds between falling and landing. In fact, the first-ever study of near-death experiences focused exclusively on falls. It was by a Swiss climber named Albert Heim and published as *Notes on Fatal Falls* in 1892. Heim fell more than 65 feet (20 metres) from a cliff, during which he felt an intense sense of well-being and calmness, a massive slowing down of time, an intense clarity of perception and had a "life-review". Over the next few years, he collected 20 examples of similar experiences from fellow climbers.

However, even if we use these wider definitions of NDEs, the experiences are certainly most common during periods of clinical death, particularly following cardiac arrests. In 2014 an international study (led by Dr Sam Parnia at the State University of New York) of more than 2,000 cardiac arrest patients was published. This found that 40 per cent reported some form of awareness during the time when they were clinically dead, when their hearts had stopped beating and their brains had shut down.[4] In a similar 2001 study by the Dutch Cardiologist Pim van Lommel – who began to investigate the experiences after so many of his patients talked about them following resuscitation – 64 out of 344 cardiac patients reported near-death experiences.[5] More generally, the research of Bruce Greyson has suggested that 10–20 per cent of patients who are close to death have the experience.[6]

So, as I told my colleague John, near-death experiences are by no means uncommon. A 2001 survey in Germany found that 4 per cent of a sample of 2,000 people had had an NDE.[7]

The aftermath of NDEs

Near-death experiences make so little sense in terms of the materialist paradigm that – as with my colleague John – people sometimes struggle to understand and accept them. People are often reluctant to tell friends or relatives about them, for fear that they will be thought of as crazy. For example, in one sample of around 100 NDEs it was found that 57 per cent of people were afraid to talk about their experiences, and 27 per cent didn't tell anyone about them for more than a year after they happened.[8] In extreme cases, people may feel so much resistance to the experience that they may completely repress it.

This was the case with another acquaintance of mine, William Murtha, whose experience I discussed at length in my book *Out of the Darkness*. William had a near-death experience while drifting in and out of consciousness in a freezing sea for around an hour. He felt himself leave his body, felt at one with everything, and became aware of what he called a "higher presence" who communicated with him. He also experienced a "life-review" of a series of the most significant experiences of his life. However, William didn't have a framework to make sense of his experience. The vision of love and connection he had experienced was also completely opposed to the individualistic and competitive values he lived by. So for the next 18 months William simply pretended that the experience hadn't happened, and he reacted by living in an even more materialistic and hedonistic way than before.

However, eventually William's resistance broke down, when he was drinking in a bar one evening and heard a voice inside his head say, "Bill, this is not who you are. You are not meant to be here."[9] He realized that he had been running away from what had happened, and in an instant everything changed. His new self emerged, and from that point on his life changed completely. He lost the need to accumulate money and consumer goods, and he felt a new enjoyment of quietness and inactivity. Shortly afterwards, he sold his share in his company and began to live a much slower and more simple lifestyle.

He felt a strong sense of empathy and compassion, and a new awareness of the wonder and beauty of everyday life.

This is a normal pattern. Once a person has accepted and integrated their NDE, they undergo a profound transformation. Indeed, one of the most striking things about near-death experiences is their long-term effect: they frequently bring about a profound shift of values and perspective, which itself leads to major lifestyle changes. People often become less materialistic and more altruistic, less self-oriented and more compassionate. They often feel a new sense of purpose and their relationships become more authentic and intimate. They report becoming more sensitive to beauty, and more appreciative of everyday things. One person who had an NDE after a heart attack told the researcher Margot Grey: "Since then, everything has been so different... The sky is so blue and the trees are much greener; everything is so much more beautiful. My senses are so much sharper."[10] People often report becoming more intuitive too, and even sometimes that they have developed psychic abilities. Another woman told Margot Grey that she felt: "a very heightened sense of love, the ability to communicate love, the ability to find joy and pleasure in the most insignificant things about me... I seemed to have a very heightened awareness, I would say almost telepathic abilities."[11]

One of the most significant effects of NDEs is a loss of fear of death. Because NDEs have such a powerful quality of realness, most people are convinced that they really have briefly experienced death. As a result, they become certain that there is life after death. And since their NDE was such a blissful experience – so blissful that people are sometimes disappointed to return to their bodies – any anxiety they may have had about dying dissolves away. It's probable that an unconscious fear of death is a major source of a lot of pathological human behaviour – such as materialism and status-seeking – so when this fear disappears, it has a major effect. So the loss of fear of death probably contributes significantly to some of the other changes I've already mentioned, such as a shift away from materialism.

It's remarkable that one single experience can have such a profound, long-lasting transformational effect. And this is illustrated by research showing that people who have near-death experiences following suicide attempts very rarely attempt to kill themselves again. This is in stark contrast to the normal pattern – in fact, a previous suicide attempt is usually the strongest predictor of actual suicide.

The fact that they have such profound after-effects makes it seem very unlikely that NDEs are a brain-generated hallucination. Hallucinations certainly do not have these kind of transformational after-effects. They are usually quickly forgotten, with a clear sense that they were delusional experiences, less authentic and reliable than ordinary consciousness. But with near-death experiences, there is a clear sense that what is experienced is more real and authentic than normal consciousness, and the person's vision of reality – and values and attitude to life – are completely transformed.

Explaining NDEs

From a materialistic point of view, near-death experiences are problematic, to say the least. Once a person's heart has stopped beating, the brain shuts down within 15–20 seconds. So how can a person continue to be conscious during this period? And how can people in a deep coma – with barely any signs of brain activity – have extremely intense and complex conscious experiences? This suggests a lack of a direct relationship between consciousness and the brain. If consciousness can occur in the absence of brain activity, then it is difficult to see how it could be the product of brain activity. It must be essentially independent of the brain, and arise from another source. One of the most fundamental principles of materialism – that matter is primary and gives rise to mind – is therefore contravened.

Because of this, materialist scientists have invested a great deal of effort into trying to explain near-death experiences in materialistic terms. So many different explanations have been suggested that it's impossible to cover them all in detail, so here we'll look at some of the most significant and evaluate their validity.

Are NDEs caused by undetected brain activity?

Many NDEs (such as those following cardiac arrest) *appear* to take place when the brain is shut down, but can we be sure that this is really the case? Perhaps there is brain activity at a very low level that is difficult to detect.

However, this argument is doubtful from a medical point of view. After cardiac arrest, brain stem reflexes are lost straight away, and do not return until the heart has been restarted. So how could a brain be functioning without showing any signs at all of activity? As Sam Parnia has noted: "It takes a lot of imagination to think that there's somehow a hidden area of your brain that comes into action when everything else isn't working."[12]

But even if there was a possibility of very low-level brain functioning, there is still a giant flaw in this argument: if the brain only has a very low level of activity in these moments, how could the complex mental functioning of a near-death experience arise? In conventional models of the mind and brain, the complex cognition, sensory perception and memory of NDEs should be impossible in a situation where brain activity is so limited as to be undetectable. It's not even just a question of people retaining a *normal* level of mental functioning in these moments – in NDEs, perception and cognition actually become more intense than normal. As we have seen, people often report feeling much more alert than normal, with a very clear and intense form of awareness. So how could a higher-than-normal level of mental activity be generated by such a low level of brain activity? It's exactly the reverse of what we would expect. If there was any conscious experience at all in these moments, it would surely be dim, vague and confused.

This argument is even more relevant to NDEs in comas. If there was a direct relationship between the brain and the mind, we would expect the conscious experience of coma patients to be very limited. Such a low level of brain activity should only give rise to very vague and confused mental experience. But again, in coma-related NDEs, the opposite seems to be the case. It's almost as if the less the brain is able to function properly, the more vivid and intense experience becomes.

This also applies to NDEs that occur under general anaesthetic. Let's return to John's experience at the beginning of this chapter. Even if we assume that the experience did not occur at a time when John's brain was shut down, it would still have had to occur at a time that he was deeply unconscious under a general anaesthetic. This applies to every single NDE that occurs in an operating theatre. It should be impossible for people who are under general anaesthetic to have any conscious experience – and particularly the very vivid and intense kind of experience that NDEs feature. As the authors of *Irreducible Mind* put it, the conventional view that conscious experience is produced by neurological activity "must be incorrect, because in both general anaesthesia and cardiac arrest, the specific neuroelectric conditions that are held to be necessary and sufficient for consciousness are abolished – and yet vivid, even heightened awareness, thinking and memory formation can still occur".[13]

It could be argued that such experiences don't actually occur during the period of unconsciousness, but as a patient is coming to, and while their anaesthetic is wearing off. But the hazy, confused return to consciousness after an anaesthetic is completely different to the alert clarity of the near-death experience. This applies to emerging from a coma too. The eminent psychiatrist Oliver Sacks suggested that Eben Alexander's NDE did not occur while he was in a coma, but "as he was surfacing from the coma and his cortex was returning to full function".[14] However, Sacks omits to mention that the normal experience of coming to after a coma is completely unlike the experience described by Alexander. In contrast to the clear, heightened consciousness that Alexander describes, mental functioning is usually very confused and impaired, taking days to return to normal. And it goes without saying that this normal, confused emergence from both anaesthetic and a coma does not have the kind of profound, transformational effect as NDEs.

Are NDEs neurological effects of the "dying brain"?

One explanation of NDEs that has been put forward many times is the "dying brain" hypothesis. This takes a similar approach to Oliver Sacks's argument, but focuses on the period preceding the loss of consciousness. According to this view, NDEs are unusual experiences that occur shortly before the brain becomes inactive. They are a kind of hallucination generated by a dying brain. In particular, this argument relies on the idea that cerebral anoxia (a lack of oxygen to brain tissue) causes many of the characteristics of NDEs. It results in "cortical disinhibition" and intense, uncontrolled brain activity. The vision of tunnels and lights could be explained in terms of disinhibition in the brain's visual cortex. At the same time, the intense sense of well-being of NDEs could be caused by the release of endorphins when a person is close to death.

However, there are serious problems with these explanations. You would expect intense, uncontrolled brain activity to result in crazy, chaotic experiences, but NDEs are usually very serene and well-integrated experiences – certainly not what one would associate with "disinhibition" and over-stimulation. As Penny Santori, the contemporary researcher of NDEs – whose perspective is informed by many years working as an intensive care nurse – points out, oxygen deficiency is associated with confusion, irritability and memory loss. The clarity and highly ordered nature of NDEs is, in her words, completely inconsistent with "a disorganized brain with no or greatly reduced blood flow".[15]

You would also expect uncontrolled brain activity to result in a very wide range of different experiences, as varied and different as dreams. However, as we have seen, the majority of people who report a continuation of consciousness in NDEs report the same "core" experience. And once again, there is absolutely no question that normal experiences of oxygen deficiency have any of the transformational effects of NDEs.

Another significant argument against the "dying brain" hypothesis is that, as we have seen, NDEs do not only occur when a person is in the process of dying. They can occur in comas or during a fall, when biologically a person may not be near death at all. The "dying brain" can't be the cause of NDEs in these situations, so it is highly unlikely that it is the case for NDEs in situations such as cardiac arrest either.

Veridical perceptions

However, perhaps the most serious problem with the dying brain hypothesis is that – as we saw with John – it is common for people to report accurate details of what they witnessed while they were apparently unconscious. This would obviously be impossible if the experience occurred as the brain is dying, and before unconsciousness. In the study of near-death experiences, these are sometimes called "veridical" – as in truthful – perceptions.

These veridical perceptions provide possibly the most convincing evidence of all that NDEs are not brain-generated hallucinations. There are many cases in which NDE patients have provided remarkably specific observations, which were later verified by the medical professionals present during procedures. There are so many well-documented examples of this that it is difficult to know which to choose. (A recent study of NDEs called *The Self Does Not Die* contains over 100 cases.) Here I'll focus on two of the most striking cases that I'm aware of.

The first is the case of a man from New England called Al Sullivan who had an emergency heart operation at the age of 56. During the operation, he became conscious and felt himself rise out of his body. He saw himself lying on a table, covered by light blue sheets. He could see the incision that had been made to reach his chest cavity, with his heart inside. His chest was being held open by metal clamps. Al recognized the surgeon he had met before the operation, who he saw – to his surprise – "flapping" his arms as if trying to fly. The surgeon was the only person near his opened chest, while two other surgeons were busy working on his leg. This confused Al, until he learned later that leg veins are often used in heart bypass surgery. After this, Al became aware of a powerful yellow light, felt intense feelings of joy and love, and saw deceased members of his family.

When he was able to speak after the operation, Al told his cardiologist, Anthony LaSala, about his experience. The cardiologist was sceptical until Al talked about the surgeon flapping his elbows. This was a peculiar habit of the surgeon, whose name was Hiroyoshi Takata. When not operating, Takata would lay his palms flat on his chest and direct his assistants by pointing with his elbows. He did this to avoid touching anything with his sterile hands. LaSala was amazed that Al had mentioned this, and told the surgeon about it.

In 1997 the NDE investigator Bruce Greyson interviewed both Anthony LaSala and Hiroyoshi Takata. Both confirmed that Takata had a habit of flapping his elbows to give instructions, and LaSala mentioned that he was not aware of any other surgeons who did this. LaSala also confirmed that Al's eyes were taped shut during the operation and that there was a drape over his head, so that he would have been unable to see the surgeon's movements even if he had been conscious. Another investigative team confirmed that Sullivan must have been deeply unconscious under anaesthetic at the time that the surgeon made the flapping movements. Later, Hiroyoshi Takata himself commented on the incident, noting that although he had heard other doctors talk about anaesthetic wearing off during operations so that patients could hear conversations, "I have never encountered one in which the patient describes such details of the operation as if he/she saw the process. Frankly, I don't know how this case can be accounted for. But since this really happened, I have to accept it as a fact."[16]

In the second case, the critical care physician Laurin Bellg reported the story of a patient called Naomi, who had had a cardiac arrest. The medical team used CPR and electric shocks to try to restart her heart, but had no success – in fact, during the procedures Naomi had a second cardiac arrest. Then the team discovered that she had a blocked coronary artery, which they unblocked. However, her heart was too weak to pump blood, which meant that her lungs didn't function properly, so she was put on artificial ventilation.

It was when Dr Bellg decided that Naomi could be taken off artificial ventilation that the patient reported a near-death experience, and described events that had taken place throughout the medical procedures she had undergone. First of all, she described scenes from the emergency room, telling Dr Bellg, "I saw everything. I saw it all... I couldn't figure it out at first, then I realized I was up above my body watching everyone rush around. I saw them pumping up and down on my chest and putting a breathing tube in my mouth. I saw my closed eyes and how limp I was with one arm hanging off the bed."[17] Then Naomi described being in a different room, with a large light overhead and different medical personnel; she mentioned a specific incident where members of the resuscitation team tilted her body and slid a long flat board under her. As they did this, she heard Dr Bellg say, "Whoa, whoa, whoa, my stuff", and watched her grab some medical items that almost fell to the floor.

Naomi was puzzled by the board, but Dr Bellg explained that its purpose was to aid the circulation of the blood to make chest compressions more effective. As Dr Bellg remarked about the case: "I remained amazed that she had been aware of that happening and saw me reacting to the shifting field by grabbing my supplies to keep them from falling off the bed when I knew, for a fact, that she was totally unconscious."[18]

It is sometimes argued that such reports could be a mental construction of what people expect to see during an operation, based on reports they have heard or television programmes they have seen. However, the reports earlier contain such specific details – many of which were completely unknown to the patients beforehand, such as the board mentioned by Naomi or the surgeons working on Al's leg – that this seems unlikely.

In fact, this hypothesis has been tested by NDE researchers. Penny Sartori asked a group of cardiac patients who had been resuscitated without having an NDE to describe what they thought were resuscitation procedures, and compared these to the descriptions of NDE patients.[19] Another researcher, Dr Michael Sabom, did the same but with a group of cardiac patients who had not been resuscitated.[20] In both cases the groups' descriptions were much less accurate and believable than those of the NDE patients. The NDE patients included many accurate and specific details, while the other groups made significant errors. Some members of the non-NDE groups had no idea at all about the resuscitation procedure, while others appeared to have a distorted picture of the procedure as a result of TV shows. Both Sabom and Sartori concluded that the accuracy of the reports of resuscitation procedures from NDE patients could not be explained in terms of expectation, prior knowledge or guesswork.

It is also worth pointing out that, as the two cases I've referenced show, the reports of NDE patients often manage to convince physicians of their veracity. In fact, one of the striking things about NDE researchers is that, in many cases, they are not parapsychologists but physicians who became convinced of the reality of the phenomenon, based on the reports of their patients. The physician mentioned in the second report, Laurin Bellg, heard so many reports of NDEs, and was so struck by the specific details that many of them contained, that she collected and published them as a book called *Near Death in the ICU*. As mentioned above, the British NDE researcher Penny Sartori worked as an intensive care nurse for 17 years, and began to research NDEs after so many of her patients reported them. The same is true of the contemporary Dutch NDE researcher, Pim van Lommel, who spent 26 years as a cardiologist. Similarly, one of the most renowned American NDE researchers, Dr Martin Sabom, is a cardiologist who specializes in resuscitation.

Other possible explanations

Before we turn towards a spiritual interpretation of NDEs, we should briefly consider other materialist explanations that have been put forward. One theory is that NDEs may be the result of paroxysm of the temporal lobes close to death, similar to the seizures of temporal lobe epilepsy. According to the theory, paroxysm may produce out-of-body experiences and feelings of ecstasy. However, temporal lobe seizures often bring perpetual distortions and feelings of loneliness and sadness, completely unlike the clarity and well-being of near-death experiences. In addition, they don't feature "core" characteristics of NDEs such as the "life-review" or encounters with deceased relatives.

There is another theory that NDEs may be due to the release of ketamine-like brain chemicals during periods of intense stress. Like temporal lobe seizures, ketamine can produce feelings of being out of the body, as well as a sense of well-being. However, studies comparing users of ketamine and survivors of NDEs have also found significant differences. The NDE groups were much more likely to communicate with deceased relatives, or to see a bright light, while the ketamine groups were much more likely to see bizarre and frightening imagery. The latter also had a strong sense that they had had an *illusory* experience – whereas, of course, the NDE groups felt that their experience had been *more* vivid than normal. Research also suggests that while a person's *first* ketamine experience may have some similarities with an NDE, with further experiences the similarities tend to fade away. Finally, there is no evidence that ketamine-like brain chemicals actually are released when a person is close to death or is undergoing intense stress.[21]

This is also an issue with the similar theory that NDEs are caused by the release of large amount of DMT – a hallucinogenic drug that our bodies naturally produce. Normally our bodies only produce tiny amounts of DMT, but the theory is that when a person is close to death, a large amount is suddenly released that creates the "psychedelic-like" experience of the NDE. However, there is no evidence that the brain releases DMT when a person is close to death. And in any case studies have shown that only a small proportion of DMT experiences (when it is taken as a drug) have any significant resemblance to NDEs.[22]

Other suggestions have been that NDEs are associated with high concentrations of carbon dioxide, altered serotonin activity or REM sleep patterns. All of these theories are problematic for similar reasons: there is an absence of some of the "core" characteristics of NDEs, a lack of a subjective feeling of reality, a lack of transformational effect and so on. In addition, all of these theories presume that the NDE experience occurs before a person loses consciousness, and so they obviously can't account for the veridical perceptions that NDEs often feature.

From a different perspective, there are some psychological theories of NDEs. One is that NDEs are due to depersonalization. In states of intense distress and anxiety – for example, during torture or rape – people sometimes report becoming disassociated from their own bodies, feeling that they have lost their sense of identity and ability to feel. This seems to be a defence mechanism; so perhaps when a person is close to death – which usually equates to a state of distress and anxiety – they may also use this strategy, resulting in the near-death experience. However, the similarities between depersonalization and the near-death experience are very limited. We have seen that NDEs feature a strong sense of clarity and alertness, which is completely unlike the confusion and unreality of depersonalization. And, again, depersonalization doesn't feature important elements of the NDE, such as a bright light or the encounter with deceased relatives.

Another psychological theory is the "expectancy model". This suggests that the NDEs are a kind of fantasy that is created as an alternative to stressful or dangerous situations – and one that is based on popular awareness of NDEs, or on religious or spiritual narratives that are similar to NDEs. However, there are some very sound reasons why this theory isn't credible. First of all, people who have no prior knowledge of NDEs describe the same type of experience as others. And when children have NDEs, they also consistently report the same characteristics as adults, even though they are too young to construct personal and cultural expectations. NDEs only became popularly known in the mid-1970s, following the pioneering investigations of Dr Raymond Moody (published in his book *Life After Life*). Indeed, the term "near-death experience" was not coined until 1975. If expectancy was a factor, one would expect NDEs that were reported after 1975 to conform more closely to Moody's model, but studies of NDEs reported before and after the publication of his book have shown no significant differences. Finally, studies have also shown that many NDEs differ markedly from people's expectations, and consequently lead to confusion and discord. John's and William's experiences, as described earlier, are examples of this. In fact, if NDEs are generated by cultural expectations, there is no reason why people like John and William – who accepted the materialistic paradigm of reality – should have NDEs at all.

NDEs from a spiritual perspective

Even academics who favour materialist explanations of NDEs admit that the theories described earlier are problematic. For example, as Harvey Irwin and Caroline Watt have put it, "It is fair to say that no current neurophysiological or psychological theory of NDEs is satisfactory."[23] In Chapter 3 I suggested that when there are so many different and divergent suggested explanations of how a phenomenon is caused (in that case, on how the brain might produce consciousness), we should suspect that the basic causal assumptions that underlie the different suggestions are flawed. And this applies even more to NDEs. The variety of different theories is bewildering, and there is almost a sense of desperation about some of them. You could compare it to a lazy school pupil who uses endless excuses to explain why they haven't done their homework – no matter how ingenuous his excuses are, the sheer number of them has the overall effect of diminishing their credibility. It's also not dissimilar to the way that fundamentalist Christians deny the evidence for evolution, or other scientific findings that cast doubt on the literal truth of the Bible – for example, by believing that fossils aren't really millions of years old but were put there by God when the world was created as a way of testing our faith. There is a determination to defend a worldview, and a refusal to countenance evidence that appears to dispute it.

The most logical way of looking at NDEs is to accept that they cannot be explained in materialistic terms. Like consciousness itself, they cannot be accounted for in terms of brain activity. But from a spiritual perspective, NDEs can be explained easily. As we have seen, consciousness is a fundamental quality of the universe, and our individual consciousness (or mind) arises when that fundamental consciousness is received and canalized by the brain. Our sense of identity is therefore not directly produced by the brain, and as a result doesn't necessarily end when the brain ceases to function. It can continue in the absence of any signs of physiological or neurological activity – and, just as importantly, it can function in very intense and integrated way when neurological functioning is at a very low level (such as in a coma). Like the material we looked at in chapters 4 and 5, NDEs point to a fundamental independence between the mind and the brain.

It is difficult to examine near-death experiences objectively and come to any other conclusion. Indeed, this is the conclusion that most NDE researchers have reached. As Penny Sartori has written, "consciousness is primary... the brain mediates rather than creates consciousness."[24] Or, in the more descriptive words of Pim van Lommel: "complete and endless consciousness is everywhere in a dimension that is not tied to time or place, where past, present and future all exist and are accessible at the same time."[25]

One potentially problematic issue here is that, according to my argument, our individual minds require the functioning of the brain in its role as a transmitter. So shouldn't our individual minds still shut down when the brain stops functioning, in the same way that a radio stops broadcasting when its circuitry is broken? Shouldn't a person who suffers a cardiac arrest still undergo a cessation of consciousness?

There are two ways of looking at this. First of all, NDEs that involve a shutting down of the brain never last more than a few minutes in real time, so one possibility is that consciousness only continues for a short period of time after brain-death, like an echo. Perhaps the individual identity that was formed by the interaction of universal consciousness with the brain can sustain itself for a while without the brain, but then slowly dissolves away. The second possibility is that once our individual minds have formed, they become permanent structures that can continue to exist without the brain.

Is there an afterlife?

The above discussion leads to an inevitable issue that arises from the discussion of NDEs – whether they offer evidence of life after death.

To many intellectuals, the idea of an afterlife is an absurdity. It is seen as a pre-scientific superstition, like the belief in fairies or witchcraft. Surely it's as irrational to believe in an afterlife as it is to believe that the world was created in seven days, or that illnesses are caused by evil spirits? Surely freedom from illusory concepts of an afterlife is one of the hard-won victories of the Enlightenment and the transcendence of religious dogma?

However, it's important to detach the idea of life after death from a religious background. The concepts of paradise or heaven in traditional religions are clearly full of fantasy and wishful thinking. It's probable that they developed – many centuries ago, when life was extremely brutal and hard for most people – as a kind of pipe dream to compensate people for the sufferings which filled their lives: a way of providing some hope in a seemingly hopeless world. But rejecting these concepts of an afterlife doesn't invalidate the idea of an afterlife altogether. It is, of course, perfectly possible to believe in an afterlife without being religious.

From the standpoint of materialism, the issue is clear: there can't be life after death because there is no such thing as a soul or spirit to survive the death of the body. Our minds are just produced by the brain, so when the brain dies, our minds die too. However, if we reject the idea that the mind is just produced by the brain – as I am doing in this book – then the idea of some form of afterlife becomes feasible.

Unfortunately, a thorough investigation of the question of life after death is beyond the scope of this book. It's such a vast subject that it requires a whole book to itself – and indeed, there are whole books on the subject. (I would recommend David Fontana's *Is There an Afterlife?*) The topic isn't wholly relevant to this book either, since my main aim is to show the limitations of the materialist view of the world, and to propose a spiritual worldview as a better alternative. But let me try to briefly summarize my perspective on the issue of an afterlife.

It's sometimes suggested that NDEs offer evidence that there is life after death. They certainly do offer evidence that consciousness isn't wholly dependent on the brain, and can continue beyond the brain and body. But as I suggested earlier, it is not clear how long this consciousness continues for. It could be that the person's identity or individual consciousness remains intact just for a brief period after death. Perhaps identity slowly dissolves away, losing its association with the essential consciousness of our being, which then becomes part of the universal consciousness again. I find that there is a satisfying circularity to this view – it's as if we are returning home, to the source from which we arose.

However, from a different perspective, NDEs actually do provide some support for the idea that personal identity remains intact, and continues indefinitely after death. This comes from the deceased relatives who are often encountered during NDEs. As we have seen, it's very common for people to encounter parents or grandparents (sometimes great-grandparents too, as with my student Max) who sometimes inform them that it's not their time to die yet, and they should return to their bodies. These could be hallucinations, of course, but – again, as with my student Max – there are cases where deceased relatives have provided information that was unknown to them that was later verified. There are also many cases where the near-death experiencer didn't know a person they encountered, and their identity was later verified by relatives. In addition, there have been cases where the experiencer encountered a relative whom they did not yet know had died.

Even by itself, this offers significant support to the idea of post-mortem survival. But there is other evidence too. In fact, supporters of the idea of post-mortem survival usually refer to three different sources of evidence: first, the evidence from mediums who appear to be able to make contact with deceased individuals; second, visions of, or encounters with, the deceased; and third, cases of apparent reincarnation.

Mediumship is widely disparaged nowadays, and it's true that there are many charlatans (and perhaps self-deluded individuals) who simply use techniques to trick gullible people into believing that they are in contact with their deceased relatives. But this should not invalidate the whole field of mediumship any more than the whole discipline of yoga is invalidated by the fact that some yoga instructors are charlatans without training. And in recent years, there have been a number of tightly controlled experiments – single-, double- and even triple-blind studies[26] – with gifted mediums, which have had significantly positive results. In one typical procedure, the experiment uses a "proxy sitter". In other words, the actual sitter is isolated from the medium; another person interacts with the medium and elicits information about the actual sitter's "discarnates" (that is, the deceased individuals linked to the sitter). The information (which often includes details such as the discarnate's appearance, personalities, hobbies and interests, and cause of death) is transcribed, and the sitter is given the transcript, together with a second transcript from a different session, where the same medium was with a different sitter. The sitter then has to simply pick the transcript that comes from their discarnate relating to them. The procedure is repeated many times over different sittings with the same medium. Many studies with this type of protocol have shown a fairly consistent – and highly significant – "strike rate" of 75 per cent.[27] As a group of researchers wrote recently, the findings of such studies strongly suggest that: "certain mediums can report accurate and specific information about the deceased loved ones (termed discarnates) of living people… without any prior knowledge about the sitters or the discarnates, in the complete absence of any sensory feedback, and without using fraud or deception."[28]

Another interesting facet of the medium-related evidence is the range of astounding abilities and creative fears which some "discarnates" have displayed. One of the most striking cases was of a "discarnate" chess grand master who played – via a medium, of course, who was not a chess player – with the living chess grandmaster Victor Korchnoi in 1985. Through the medium the discarnate identified himself as a Hungarian grand master, Geza Maroczy, a well-known player who died in 1950 and was ranked third in the world in his prime. The game continued for 48 moves, and analysis by chess experts verified that Maroczy's abilities were at grandmaster level (although he lost). Experts also attested that his style of play was rather old-fashioned, and consistent with Maroczy's actual style. In addition, the "discarnate" Maroczy was asked 81 questions about the life of the "actual" Maroczy, and 79 of the answers could be authenticated as accurate.[29]

There are a variety of types of visions of, or encounters with, the deceased. One example is the encounters with relatives during NDEs; another is "death bed visions", when dying people appear to see deceased relatives. Nurses and other caregivers often hear them describing the relatives, and calling or talking to them. When Penny Sartori started to work as an intensive care nurse, she was surprised to find that her fellow nurses used this as a sign that death was near: "As my nursing career progressed I realized that patients calling out or talking and gesturing to unseen people symbolized fast-approaching death and was commonly accepted by many nurses that I worked with."[30] Caregivers usually distinguish these episodes from drug-induced hallucinations, noting that they often occur when the patient's consciousness is clear, and that, unlike hallucinations (which normally bring anxiety and agitation), they normally lead to a sense of calmness and acceptance of death. Research by Osis and Haraldsson showed that patients were less likely to have deathbed visions if they were medicated, or if their consciousness was impaired by illness.[31] And as with encounters with deceased relatives during NDEs, there are cases of patients who encounter a deceased person whose recent death they were unaware of.

In everyday situations, it's also not uncommon for recently deceased people to appear to relatives before the latter were aware of their death. A friend of mine woke up suddenly in the middle of the night and saw her grandmother (who lived 150 miles [240 kilometres] away) standing at the bottom of the bed. She was shocked and said to her, "What are you doing here, Gran?" Her grandmother didn't answer, and then disappeared. Later that morning, my friend received a phone call to say that her grandmother had died the previous night. Other variants of these encounters including smelling the perfume of a deceased person, feeling their touch and hearing their voice, and even receiving communications from them through electronic devices, such as computers, radios or telephones. Studies have found that between two-thirds and three-quarters of bereaved partners feel that they have sensed the presence of their deceased spouse.[32]

As far as reincarnation is concerned, there are many well-documented cases of young children who have reported specific details of a past life that were later verified by investigators. The most prominent researcher in this area was Dr Ian Stevenson, a psychiatrist at the University of Virginia School of Medicine, who spent much of his life collecting and examining such cases. Typically, between the ages of two and four, such children would start talking about their previous life, often speaking about the events that led up to their death, and sometimes using the present tense, as if their previous life was still continuing. In some cases, Stevenson was able to identity the person the child claimed to be and to verify the information by speaking to relatives of the deceased.

Some of the most striking evidence examined by Stevenson related to pigmented birthmarks. A significant number of children who claimed to remember past lives had birthmarks that they explained in terms of injuries from their previous life. In 49 cases Stevenson managed to obtain medical documents of the deceased person, and in 43 of these there was information that confirmed a correspondence between the wounds and birthmarks.[33]

I believe that any impartial examination of the evidence would make it clear that it is far from irrational to be open to the possibility of some form of life after death. As the authors of *Irreducible Mind* put it: "if other things were anywhere near equal, most rational persons would conclude, on the basis of the available evidence, that survival in some form is at least possible, and perhaps even a demonstrated empirical reality."[34] The authors also note that, unfortunately, "other things" are nowhere near equal, due to the dominance of the materialist paradigm.

As I noted earlier, for me the issue is not whether there is an afterlife or not, but *how long* individual identity continues after death. And the evidence strongly suggests that identity continues indefinitely.

NDEs as spiritual experiences

To conclude this chapter, let's briefly return to NDEs. Their real importance in terms of the argument of this book is that they provide convincing evidence that consciousness is not directly produced by the brain, but is rather an essential quality of the universe. And it is important to consider that this is indicated by the content of NDEs, as well as their occurrence itself.

In NDEs it is common for people to experience a sense of oneness – a feeling that all things are interconnected, and that they are part of this interconnection too. People also sometimes report an awareness that the fundamental reality of the world is love, or that there is a kind of force pervading everything, with qualities of radiance and bliss. In these moments it may be that they are directly experiencing spirit-force, as mystics and indigenous people were able to (as described in Chapter 2). The oneness they experience is the shared essence of all things, with spirit-force pervading them. The translucent light they see is a quality of radiance that spirit-force innately possesses.

In this sense, NDEs are significant because they are, in part, powerful *spiritual* experiences. We saw earlier that NDEs have profound transformational effects, and in my view this is due to three different factors. The first factor is that, irrespective of their content, NDEs involve a close brush with death. They are a powerful reminder of the reality of death, and the fragility and the temporary nature of life, and as a result they bring about a new perspective and a new sense of appreciation. The second transformational factor is that NDEs alleviate the *fear* of death, partly because they show that the experience of dying can be euphoric and liberating and also because they convince people that there is life after death. And the third – and perhaps most important – transformational factor is the spiritual content of NDEs. The vision of love, beauty and connection that people perceive – and the sense of euphoria and oneness they experience – changes them permanently. They sense that they have seen the essential truth of things, a more intense and complete vision of reality than our normal awareness permits us. Although the initial intensity of that vision may fade, something of it remains with them, and as a result they never see the world or life in the same way again. They undergo an experience of spiritual awakening.

And this leads us elegantly into the next chapter, where we will discuss spiritual experiences in more detail.

CHAPTER 7
WAKING UP: THE PUZZLE OF AWAKENING EXPERIENCES

A few years ago, when I was on holiday with my family in Wales, I decided to explore the farmland around our rented bungalow. I climbed over a gate and found myself looking down at a valley, with farmers' fields sloping as far as I could see and hundreds of sheep dotting the hills. After I'd been walking for a few minutes, looking at the fields and the sky, there was a shift in my perception, as if someone had pressed a switch. Everything around me became intensely real. The fields and the bushes and trees and the clouds seemed to be powerfully there, as if an extra dimension had been added to them. They seemed more vivid, more intricate and beautiful. I also felt somehow connected with my surroundings. As I looked up at the sky, I felt that somehow the space that fills it was the same "space" that filled my own being. What was inside me, as my own consciousness, was also "out there". There was a glow of intense well-being inside me, too, a powerful feeling that "all is well". It felt incredibly "right" to be alive in the world.

This is an example of what I call an "awakening experience". As a psychologist, I have been studying such experiences for a decade. I define an awakening experience as a temporary expansion and intensification of awareness that brings significant perceptual, affective and conceptual changes. According to my research, the three most common characteristics of the experiences are: heightened awareness, positive affective states (including a sense of elation or serenity, a lack of fear and anxiety, and a sense of appreciation) and a sense of connection (towards other people, nature or the whole world in general). The latter characteristic involves a transcendence of separateness. When this occurs, we no longer feel as if we're "someone" who lives wholly inside our own mind or body, with the rest of the world "out there", on the other side. Other significant characteristics include love and compassion, a sense of being intensely present and a sense of inner quietness or emptiness.

Awakening experiences carry a strong sense of revelation and conviction. As with NDEs, there is a strong feeling that, rather than being hallucinations or delusions, they are more real than normal awareness. We feel that filters and obstacles that restrict our normal awareness have fallen away, and so we are seeing a wider and truer vision of the world. It's as if we're seeing in three dimensions rather than two, or in colour rather than in black and white. In comparison, our normal awareness seems incomplete and even unreal. There is a sense of, "Yes, so this is the way things *really* are!"

When the experience ends, there may be a feeling that we have fallen back "asleep" – that is, that our awareness has become restricted and limited again. However, the knowledge we have gained – including the awareness that this expanded reality is there, and is therefore always potentially accessible to us – may give us comfort, and a new sense of trust and optimism. As one participant in my research told me, "That moment allowed me a glance into the other side and opened me to the knowing that I am never separate." Or as another person stated, "To know that it's there (or here, I should say) is a great liberation."

Outside religion and spirituality

Awakening experiences have been consistently reported throughout recorded history, across different cultures. In the past, they were – and often still are – referred to as spiritual or mystical experiences, and associated with both Eastern spiritual traditions and Western mystical traditions. However, in the modern world they usually occur outside the context of these traditions, and without being induced by spiritual practices. The great majority occur unexpectedly in the midst of everyday activities and situations, to people who know little or nothing about spirituality or religion. That's one reason why, rather than the more traditional terms, I prefer to use the term "awakening experience", emphasizing that the experience is a completely natural psychological phenomenon that can be interpreted in spiritual or religious terms, but doesn't have to be.

Research suggests that awakening experiences are common. A 2008 study found that 80 per cent of 487 respondents claimed to have had profound ecstatic experiences of a religious, mystical or aesthetic nature.[1] However, as with NDEs, there seems to be some reluctance to openly discuss awakening experiences. Because they don't fit easily within the materialist paradigm, and because they are traditionally associated with spirituality and religion, awakening experiences are slightly taboo. People are afraid of talking about them for fear of being thought of as crazy, or as woolly minded and irrational.

A couple of years ago I gave a talk about transpersonal psychology in a bookshop, and mentioned my research into awakening experiences. At the end of the evening, a woman came up to me, looking slightly confused. "You know the experiences you described earlier?" she asked. "I'm an atheist, but I had an experience like that. Is that possible?"

"Yes!" I replied. "Awakening experiences don't have to be connected to religion, or even with spirituality. If a religious person has an awakening experience, they'll probably interpret it in religious terms. But non-religious people have them too, and interpret them in a different way. Fundamentally, it's just an expanded state of awareness, which anyone can experience."

She went on to describe an awakening experience she had while travelling on a train, looking out of the window. Suddenly there was a feeling of liberation and clarity. Time stopped and everything she looked at seemed alive and beautiful. She was filled with a sense of serenity, a feeling that everything was in harmony. She seemed intensely relieved that she could make sense of this experience, and that it didn't have to be explained in religious terms. Like so many people in our culture, it seemed that she believed there were only two metaphysical alternatives – either religion or scientific materialism – and that any experiences or phenomena that don't accord with materialism had to be categorized as religious.

The triggers of awakening experiences

Sometimes awakening experiences seem to occur spontaneously, for no apparent reason. However, in most cases they seem to be associated with certain triggers, or activities or situations. In my research, I have found that there are three major triggers that consistently show up, as well as a host of less significant ones.

The most common trigger of awakening experiences may seem puzzling initially: psychological turmoil. I have found that around one-quarter of awakening experiences occur in situations of stress, depression and loss. For example, a woman described how she was devastated by the end of a seven-year relationship. However, in the midst of this suffering, she "began to experience a clearness and connection with everything that existed… I was in a state of such pure happiness and acceptance, that I was no longer afraid of anything. Out of that depth arose such a compassion and connection to everything that surrounded me that I could feel even the pain of the flowers being picked."[2]

An American man reported an awakening experience that occurred in combat in Vietnam. He was carrying a wounded marine to a waiting helicopter when: "something happened to me. It is actually indescribable but I will make a feeble attempt to do so. I opened up, literally, from my perspective. I came out of myself. I expanded infinitely. I disappeared. It didn't last long but it was the most powerful experience I've ever had."[3] Another person described how he went through a long period of inner turmoil due to confusion about his sexuality, which led to the breakdown of his marriage. But again, in the midst of this turmoil, while on a final family holiday in Tunisia, he had an awakening experience in which: "Everything just ceased to be. I lost all sense of time. I lost myself. I had a feeling of being totally at one with nature, with a massive sense of peace. I was a part of the scene. There was no 'me' anymore."[4]

The second major trigger of awakening experiences according to my research is contact with nature. As with my experience at the beginning of this chapter, people frequently report awakening experiences that occurred while they were walking in the countryside, swimming in lakes or gazing at beautiful flowers or sunsets. For example, one woman reported an awakening experience that occurred when she was swimming in a lake, when she "felt completely alone, but part of everything. I felt at peace… All my troubles disappeared and I felt in harmony with nature. It only lasted a few minutes but I remember a sense of calmness and stillness and it soothes me now."[5] These are the type of ecstatic experiences of connection to nature that are often described by Romantic poets such as William Wordsworth and Percy Shelley.

The third trigger of awakening experiences according to my research is spiritual practice. This primarily means meditation, but also includes prayer and psycho-physical practices such as yoga or tai chi. The relaxing, mind-quietening effect of these practices seems to facilitate awakening experiences.

After these three triggers, there are several slightly less significant ones, including watching or listening to arts performances, reading (particularly spiritual literature), creative activities (such as dancing or playing music), love and sex. Only a small number of sexual awakening experiences were reported to me, but it is likely that participants were reluctant to divulge such intimate experiences, so that in reality the frequency of such experiences may be higher. Curiously, although awakening experiences are sometimes popularly associated with psychedelic drugs – such as LSD, "magic mushrooms" or ayahuasca – this has not been substantiated by my research. For example, in a study I conducted of

161 awakening experiences, only eight occurred under the influence of psychedelics.[6]

Different degrees and types of awakening experiences

There are different intensities of awakening experiences, with different characteristics that emerge at these different intensities. A low-intensity awakening experience may just feature a sense of heightened awareness, that our surroundings have become more real, with qualities of "is-ness" and "alive-ness". Medium-intensity awakening experiences may feature a sense that all things are interconnected, as if they are expressions of something more fundamental, and so share the same essence. We usually feel that we are part of this oneness too, so that we lose the sense of being separate and isolated individuals. We may also feel a strong sense of compassion and love for other people, recognizing that we share the same essence of them.

In high-intensity awakening experiences, the whole material world may dissolve away into an ocean of blissful radiant spirit-force, which we feel is the essence of both the universe and our own being; we may feel that we literally *are* the universe. This is equivalent to what yoga philosophy refers to as *nirvikalpa samadhi*, in which awareness expands beyond the boundaries of the normal self and we lose the sense of being an "I" altogether. We don't just become one with the absolute reality, we actually *become* it. In Christian mysticism, the 13th-century German mystic Meister Eckhart described this in terms of union with the "Godhead", the unconditioned source from which the whole world – including God himself – flows out.

My research has shown that there is an inverse relationship between how intense awakening experiences are and how common they are. In other words, the more intense they become, the less common they are. In a recent study of 90 awakening experiences, my co-author and I found there were only 11 "high-intensity" ones. These were experiences in which time and space and personal identity dissolved away, leaving

only a sense of merging with and becoming the universe.
One person described having such an experience when he was
a young child, running out of the front door of his house:
"Everything just melted. I looked at the tarry telegraph
pole outside of my friend's house four doors up. It was just
pulsating with life and energy; the road surface was the same.
I looked to myself, I was made up of the same pulsating
energy. Time just melted as well."[7] Another person described
how she was "in and of the universe, with time and space
altered. I knew I could be everywhere all at once. There was no
concept of distance or past and present... The sense of peace,
blissful and oneness is hard to put into words."[8]

Aside from the question of the intensity of awakening
experiences, there is an important distinction that was originally
made by the British philosopher and scholar of mysticism,
Walter Stace. This distinction is between extravertive and
introvertive mystical experiences. As their name suggests,
extravertive experiences are focused outwards. They involve
a different vision of the world, and a different relationship to
nature and other people. They may bring intensified perception,
and a sense of the oneness of things. All of the experiences we've
looked at so far in this chapter are of this type.

In introvertive experiences we turn our attention in the
other direction – into our own being, where we might feel
that we've connected to a deeper and more authentic part of
ourselves, and feel a sense of inner expansiveness, together with
a feeling of bliss and wholeness. While extrovertive experiences
often happen spontaneously, introvertive experiences are most
familiar to meditators and contemplatives, who are experienced
in withdrawing into their mental space and attaining a state of
inner quietness. At the higher levels of intensity, an introvertive
experience becomes what the philosopher Robert Forman has
called the "pure consciousness event".[9] This is when our minds
become completely quiescent and empty and we lose any sense
of individual identity. We become aware that the spiritual
essence of our being is the same spiritual essence of the whole
of reality and feel an indescribable sense of bliss and freedom.

One of the Indian *Upanishads*, the *Mandukya Upanishad*, describes it as a state that is: "Beyond the senses, beyond the understanding, beyond all expression… It is our unitary consciousness, wherein awareness of the world and multiplicity is completely obliterated. It is ineffable peace. It is the supreme good. It is One without a second. It is the Self."[10]

However, it is important to note that, ultimately, there's really no distinction between introvertive and extrovertive experiences. In the latter, we experience oneness through the world of form; we sense an essential spirit-force pervading all things and bringing them into unity. In introvertive experiences, we experience oneness through our own being, by encountering spirit-force as our own essence, and realizing that it is also the essence of everything else which exists. It's like arriving at the same destination from a different starting point. The significant point is that, at the highest intensity of awakening experiences, notions of inner and outer (or introvertive and extrovertive) don't have any meaning.

The after-effects of awakening experiences

I've already highlighted some similarities between awakening experiences and near-death experiences – namely, that people are reluctant to talk about them and also that they carry a strong element of conviction of revelation, appearing to be more real than ordinary awareness. In addition, the after-effects of awakening experiences are very similar to – although perhaps not quite as powerful as – near-death experiences.

Even though they are typically of a very short duration – from a few moments to a few hours – awakening experiences frequently have a life-changing effect. Many people describe awakening experiences as the most significant moment of their lives, bringing about a shift in their perspective on life and their values. In our recent study of 90 awakening experiences, we found that the most significant after-effect was a greater sense of trust, confidence and optimism. One person reported that even though "that whole experience was brief, it left a little piece of knowing and hope… [I]t left me knowing that your inner truth is always there for you."[11] One person had a powerful awakening experience while suffering from intense depression. The experience only lasted for a few minutes, but in its aftermath she found that the feeling of dread had disappeared from her stomach and she felt able to cope again, which led to a new, positive phase in her life. As she described it: "I looked around and thought about all the good things in my life and the future. I felt more positive and resilient."[12] Such changes in attitude sometimes led to significant lifestyle changes, such as new interests, new relationships and new careers.

The great psychologist Abraham Maslow didn't specifically study awakening experiences, but instead included them within the wider category of "peak experiences". But what Maslow wrote of peak experiences in general certainly applies to awakening experiences: "My feeling is that if it [the peak experience] were never to happen again, the power of the experience would permanently affect the attitude towards life. A single glimpse of heaven is enough to confirm its existence."[13]

It's not surprising that such similarities exist between awakening experiences and NDEs because – as we saw at the end of the last chapter – to a large extent they are actually the *same* experience. Or to put it more specifically, NDEs usually *include* high-intensity awakening experiences, along with their other features (such as an out-of-body experience, encounters with deceased relatives, a "life-review" and so on). The intense well-being, the sense of the interconnectedness of everything and the sense of oneness with the world, which NDEs feature, are exactly the same qualities that feature in awakening experiences. In other words, when consciousness continues after the shutting down of the brain, it seems to undergo the same expansion and intensification that occurs during awakening experiences. (We'll look at this in more detail shortly, especially in relation to the radiance or translucence that is often perceived both in high-intensity awakening experiences and near-death experiences.)

Materialistic interpretations of awakening experiences

If you accept the materialist metaphysical paradigm, then awakening experiences have to be explained in terms of neurological activity. And not only that, they have to be explained in terms of *aberrational* neurological activity. One of the characteristics of materialism is a naive faith in the objectivity and reliability of ordinary awareness. There is an assumption that, in an ordinary state of awareness, we see the world more or less *as it is*, and that non-ordinary (or altered) states of awareness can only be inauthentic and the vision of the world they show us can only be illusory. You could call this the hegemony of ordinary awareness – the assumption that ordinary awareness is superior to any other form of awareness. From this point of view, awakening experiences are not interpreted as more expansive states of awareness, but as hallucinations.

There have certainly been attempts to explain some aspects of awakening experiences in these terms. Some neuroscientists – such as Michael Persinger and VS Ramachandran – have suggested that mystical experiences are the result of stimulation of the temporal lobes of the brain.[14] Persinger has even claimed to be able to induce mystical experiences with a "helmet" – now called the "Shakti helmet" and available to buy – that stimulates a person's frontal lobes with magnetic fields. Another theory, put forward by the physician Andrew Newberg, is that mystical experiences of oneness occur when the part of the brain responsible for creating our sense of being an individual and boundaried self (the superior posterior parietal cortex) becomes less active than normal.[15] (In fairness to Newberg, he doesn't actually claim that spiritual experiences are *caused* by this neurological change, only that there is a correlation. Nevertheless, others have interpreted his theory in causal terms.)

However, these theories have similar problems to the attempts to explain NDEs in materialist terms. In fact, the link between the brain areas described earlier and awakening experiences is even more tenuous than neurological explanations of NDEs. One of the alleged pieces of evidence for the temporal lobe theory is that temporal lobe patients frequently have mystical experiences, but this is very dubious. As we saw in the last chapter, temporal lobe seizures are much more likely to bring feelings of anxiety and disorientation rather than spiritual experiences. And living with temporal lobe epilepsy as a long-term condition is usually a fraught and painful existence – the typical "temporal lobe personality" experiences manic tendencies, depression, altered sexuality, anger, hostility and many other negative characteristics.[16] These are so far away from the positive characteristics of awakening experiences that it's difficult to see how there could be any relationship between them. Furthermore, studies of temporal lobe patients have cast doubt on the idea that they are unusually religious or spiritually inclined. In fact, the studies suggest that their frequency of religious experience is between 1 per cent and 4 per cent, which is incredibly low compared to the general population.[17]

Apart from all this, there is absolutely no evidence that the temporal lobes do become stimulated in mystical or spiritual experiences. A very small proportion of awakening experiences do seem to be associated with physiological changes (in one of my earlier books, *Waking From Sleep*, I refer to these as instances of "homeostasis-disruption"), but this is most frequently due to fasting, sleep deprivation, self-inflicted pain or psychedelic drugs.

Another issue is that, if we look closely at the kind of experiences described by Persinger and Ramachandran, they have very little in common with awakening experiences or spiritual or mystical experiences. Ramachandran is actually describing "supernatural" experiences, such as shamanic-type journeys to other worlds or supposed regressions to previous lives. Persinger claims that his helmet produces a sense that another person or being is present, or induces visions of bright geometrical patterns. But apart from the strong feelings of well-being and a feeling of alertness, these reports don't match any of the characteristics of awakening experiences which we've discussed.

As if these issues weren't enough, these theories are also very dubious from the standpoint of scientific research. In a review of the research that relates spiritual or religious experiences to brain activity, the experimental psychologist Craig Aaen-Stockdale found numerous flaws, particularly a lack of control groups and successful replication. He concluded: "Sceptics are, in my opinion, far too quick to claim that God is 'all in the brain' (usually the temporal lobe) when in fact the evidence base is disturbingly weak."[18] Another issue in relation to Andrew Newberg's research in particular is that his claim that our sense of self – and our awareness of boundaries – stems from the posterior parietal lobe. Other neuroscientists have been very critical of this claim, and the consensus is that our awareness of boundaries is actually associated with the temporal lobe and our sense of self is mainly associated with the frontal lobe.[19]

So we have to conclude that attempts to explain awakening (or spiritual/mystical) experiences in neuroscientific terms are even more problematic than similar attempts to explain NDEs – so much so that, as they presently stand, they can be completely discounted.

Panspiritist interpretations

It could be that more convincing neurological explanations of awakening experiences will appear in the future. However, there is no reason for us to wait for these because awakening experiences can be explained easily in panspiritist terms.

The feeling many people have during awakening experiences of filters or restrictions falling away is, I believe, the key to understanding the experiences. From a psychological point of view, this is exactly what is happening in these moments. In Chapter 1 I suggested that the two most significant psychological features of our normal "sleep" state of being are our strong sense of individuality and separateness, and our automatic, familiarized perception of the phenomenal world. However, in awakening experiences all of this changes. We lose our sense of separateness and become connected to other human beings, to nature and the whole world. In addition, our perception becomes energized and "de-automatized", so that the world around us becomes more vivid, beautiful and intricate.

This shift is due to psychological changes, not neurological ones. Many awakening experiences are related to states of relaxation and mental quietness – for example, when they are induced by contact with nature, spiritual practices, creative acts or watching arts performances. These activities induce what I have called a state of "intensified and stilled life-energy". We reduce the amount of energy we expend through mental functions such as cognition, concentration and perception.

For example, consider what happens when you sit down to meditate. You remove yourself from activity and external stimuli, and then attempt to quieten your mental chatter by focusing your attention (for example, on a mantra or your breathing). As a result, your mental energy isn't being expended as much as usual, and it begins to intensify inside you. The most important aspect is quietening your thought-chatter, which is normally so incessant and turbulent that it uses up a massive amount of energy. Contact with nature can be a little like meditation, too, in that natural scenery provides a focus for our attention and a retreat from busyness, that quietens our minds and intensifies our inner energy.

The intensification of inner energy is so important because it means that we don't have to perceive our surroundings automatically anymore. Our normal automatic perception is essentially an energy-saving measure. We seem to have a psychological mechanism that "switches off" our attention to the is-ness of our surroundings, so that we don't expend energy perceiving them and so that energy can be saved for other functions. However, when our inner energy is intensified there is no need for this mechanism to function. Our perception becomes "de-automatized" and our surroundings become vivid and beautiful. We "wake up" from our normal perpetual sleep. At the same time, now that our minds are quiet and still, our normal sense of separation from the world fades away.

This sense of separation is largely maintained by our incessant thought-chatter, so when this chatter fades away, separation fades away too. (See my book *Waking From Sleep* for a fuller explanation of these processes.)

But how can we explain awakening experiences linked to psychological turmoil? It's difficult to interpret these in terms of relaxation and mental quietness, since they usually occur in states of mental agitation. This is a complex issue, which I don't have space to explain in detail, but the main point is that when people are in psychological turmoil it is often because their psychological attachments have been broken down. In other words, the external things they depend on for their sense of well-being and identity – such as hopes, ambitions, beliefs, possessions, status and relationships – have been taken away. Since these attachments build up our normal sense of identity, we temporarily experience a loss of self, or what I call "ego-dissolution". This causes distress, but it is also potentially liberating, creating a new inner openness and fullness. Our normal ego-boundaries fade away, so that we transcend separateness. And since psychological attachments normally consume a lot of our inner energy, we also experience an intensification of life-energy and a de-automatization of perception. (See my book *Out of the Darkness* for a fuller discussion.)

One way to look at our normal state of "sleep" is to see it as a state of alienation from spirit-force. Unlike earlier peoples, and indigenous peoples, we aren't able to sense spirit-force pervading the world, making all natural (and man-made) things alive and sacred. And we aren't able to sense our own connection with spirit-force; instead, we experience ourselves as separate, autonomous individuals, set apart from the natural world. (As we saw in Chapter 1, both of these factors cause our reckless and abusive treatment of the natural world: that is, we can't sense its sacredness and we don't feel connected to it.) However, when we undergo the psychological shift described earlier we are no longer alienated from spirit-force. The psychological structures that alienate us from it are undone – and so we encounter spirit-force and experience its effects.

The more intense the experience is – that is, the more fully our normal restrictions fall away – the more full and direct is our encounter with spirit-force. At lower intensities of awakening, we may not directly experience spirit-force, but are able to sense its qualities of bliss, harmony and radiance filling our own being and the world itself. We may not directly sense the essential oneness of all things, but we may feel a strong sense of connection to other people and to nature, and become aware of the interconnectedness of different phenomena. You could compare this to the sun on a cloudy day. You can't see the sun directly, but you can still see its light and feel its heat.

At the higher intensities of awakening, we experience spirit-force very fully and directly – or, in terms of the metaphor I've just used, we encounter the sun itself. We directly experience the fundamental oneness of the universe, underlying and pervading everything – including our own being – and filling the world with radiance. We directly experience consciousness as a fundamental quality of the universe. This is especially the case with introvertive experiences of pure consciousness. We find that, at the essence of our being, we are spirit-force, and so become one with the essence of the universe. We become limitless and experience an indescribable sense of bliss. (In this regard, it's important to remember that, strictly speaking, there is nothing transcendent about spirit-force. Although I've spoken about an "encounter" with it, this doesn't mean that it is something *other* to us. In a sense, it's an encounter with our own being – which is ultimately indistinguishable from the being of everything else, and of the whole universe.)

The nature of spirit-force

One of the significant things about awakening experiences is that because they are an encounter (to a greater or lesser degree) with spirit-force, they reveal to us certain essential qualities or characteristics of spirit-force. For example, awakening experiences suggest that spirit-force has natural qualities of bliss, harmony or love, which we can sense both within us and as pervading the world. This is also evident from reports of near-death experiences – as we have seen, an intense well-being and awareness of a universal quality of love are common features of NDEs.

In the Indian *Upanishad*s, we are told that the nature of *brahman* – and therefore of reality itself – is *satchitananda*: "being–consciousness–bliss". Awakening experiences – and near-death experiences – confirm this. Bliss is the nature of spirit-force in the same way that wetness is a quality of water.

Both awakening experiences and NDEs also strongly suggest that spirit-force has a natural quality of radiance. In lower-intensity awakening experiences this radiance is evident in the vividness and freshness of all things, as if they have been illuminated. In higher-intensity experiences people sometimes describe being aware of a translucent light as a quality in itself, and can feel it permeating their own being and the whole of their surroundings. (The analogy to the sun and its heat and light is even more applicable here!) This can happen in both introvertive and extravertive experiences – that is, the light can appear as both a quality of our own inner being and of the world itself. (Again, this is exactly what we would expect, since spirit-force is the essence of both our own being and the world, so that ultimately there is no distinction between them.)

In an (unpublished) example from my own collection of awakening experiences, one person described an experience of inner light as he was falling asleep one night, after spending the evening reading a spiritual book. He described how he "was suddenly overwhelmed by a spiritual light engulfing my inner being which gave me a sense of profound bliss". Another person described how she was in a depressed mood, when suddenly "there was a split in the dark inside me and the brightest white light shone out. It was like greeting a long lost friend I had forgotten but knew I knew... [I] then realized it was not a friend but ME." In extravertive terms, a person described an experience that occurred while sitting in her local park, when: "All of a sudden the leaves, trees, the whole surroundings became light… The light was touching everything, everything was interlocked/related. It was amazing, indescribable."

These are very similar to the descriptions of light in near-death experiences. It's also very significant that, across spiritual traditions, the ground reality of the universe – pure consciousness, *brahman*, the Christian mystical Godhead, or *dharmakaya* – is described as possessing a brilliant radiance. In Vedanta philosophy *brahman* is often compared to the sun: the *Bhagavad Gita* states: "If the light of a thousand suns suddenly arose in the sky, that splendour might be compared to the radiance of the Supreme Spirit."[20] Similarly, the Jewish mystical text the *Zohar* describes the universe as pervaded with translucent light (in fact, *Zohar* can be translated as "splendour" or "brilliance"). Spiritual traditions also describe our inner being in these terms. In Hindu terms, the light of *brahman* manifests itself inside us as the inner light of *atman*. As the *Katha Upanishad* states: "The light of the Atman, the Spirit, is invisible, concealed in all beings. It is seen by the seers of the subtle, when their vision is keen and clear."[21] So it seems likely that this translucent light is an innate quality of spirit-force, which we experience more powerfully and directly through increasing intensities of awakening.

We have also seen that high-intensity awakening experiences – and also near-death experiences – have a quality of timelessness. Somehow the whole of time seems to collapse into the present moment; the future and the past seem to become part of now. As Meister Eckhart put it, at the essence of our soul there is: "no yesterday or day before, no morrow of day after (for in eternity there is no yesterday or morrow); there is only a present now; the happenings of a thousand years ago, a thousand years to come, are there in the present and the antipodes the same as here."[22] This suggests that spirit-force has a quality of timelessness, and that notions of the past, present and future are constructs of the human mind.

Conclusion

Before we end this chapter, it's important to mention that although we've been discussing temporary awakening experiences, awakening can become a permanent, ongoing state too. Indeed, the primary purpose of all spiritual and mystical traditions is to "wake up" on a permanent basis. This is the aim of the systems of self-development associated with spiritual traditions, such as the eightfold path of Buddhism, the eight-limbed path of yoga, the paths of Sufism and the Kabbalah, and of Christian monasticism. All of these paths contain practices and lifestyle guidelines designed to transform our state of being by undoing the psychological structures that normally restrict our awareness – more precisely, by undoing our strongly boundaried ego, and the "desensitizing" mechanism that makes our perception automatic. All spiritual traditions have terms for this state of permanent wakefulness: in Hinduism, this is *sahaja samadhi*; in Sufism, it's *baqa* (literally, "abiding in God"); in Taoism, it's *ming*; in the Jewish Kabbalah it's *devekut*; while in Christian mysticism it is sometimes referred to as deification or theosis; and so on. There are certainly some differences in these conceptions – different traditions tend to emphasize different aspects of wakefulness, and to interpret them slightly differently according to their metaphysical and cultural constructs – but the essential principles are the same. They all refer to an ongoing state of more intense and more expansive awareness, in which the characteristics of temporary awakening experiences are established on a stable basis.

However, as with awakening experiences, this state is by no means just confined to spiritual or mystical traditions. In my research, I have found that it's not uncommon for a permanent shift into a more expansive awareness to occur in people who know nothing about spirituality. This happens most frequently in situations of intense psychological turmoil. In my book *Out of the Darkness*, I collected many examples of this from people who had suffered intense stress and depression, been diagnosed with cancer, experienced bereavement or addiction, and so on.

Many spiritual teachers – such as Jiddu Krishnamurti, Eckhart Tolle and Byron Katie – also went through periods of intense turmoil before undergoing a sudden moment of awakening. In other words, as well as being the most frequent trigger of temporary awakening experiences, intense psychological turmoil can also trigger permanent wakefulness.

You could say that everyone who has undergone this shift has transcended present-day human beings' normal state of alienation from spirit-force and is now living in constant connection to – and awareness of – it.

CHAPTER 8
KEEPING THE ACCOUNT OPEN: THE PUZZLE OF PSYCHIC PHENOMENA

To many modern academics and intellectuals, belief in psychic phenomena like telepathy, precognition and clairvoyance holds the same status as belief in an afterlife – or in fairies or alien abductions. Surely – so many of our peers seem to feel – we should have progressed beyond such irrational nonsense? Surely this was another of the hard-won victories of the Enlightenment: a rational understanding of the world, which means that we no longer have to resort to supernatural explanations? Such explanations were born out of ignorance, and now we no longer have to be ignorant. If we did accept the reality of psychic phenomena, it would mean returning to a world of superstition and witchcraft, where nothing is random, and illnesses and misfortune are caused by evil spirits. The *Diagnostic and Statistical Manual of Mental Disorders* – the handbook that American psychiatrists use to diagnose the conditions of their patients – even lists belief in psychic phenomena such as clairvoyance and telepathy as a sign of a schizotypal personality disorder.[1]

In this chapter, I hope to convince you that this view of psychic phenomena is misguided. It is not at all irrational to accept the existence of psychic phenomena – indeed, I will argue that there is so much evidence for their existence and such a sound theoretical basis for them that it is actually irrational *not* to accept their existence. It is only from the perspective of materialism that psychic phenomena appear to be impossible. From the panspiritist perspective, there is nothing anomalous about them at all.

Let me begin by clarifying what I mean by psychic phenomena (or psi, as the field of study is often termed in short), which I'll do by making a distinction. It is useful to think in terms of soft and hard paranormal phenomena. There are some types of paranormal phenomena whose existence is more questionable than others, because there isn't a lot of evidence for them – for example, UFOs, alien abductions, fairies, haunted houses, demonic possession or witchcraft – and these could be termed *soft* paranormal phenomena. On the other hand, there are types of paranormal phenomena that have been rigorously tested over many decades, and for which we have accumulated a lot of evidence. Their existence also makes sense from a theoretical point of view, especially in terms of some of the findings of quantum physics. These are what I call *hard* paranormal phenomena. This category is mainly phenomena that are sometimes referred to under the umbrella term of extra-sensory perception, or ESP, such as telepathy, precognition and clairvoyance (or remote viewing). I would also class psychokinesis – the ability to influence physical systems with mental intention, sometimes called "remote influence" – as a hard paranormal phenomenon.

Hard and soft paranormal phenomena are often lumped together, which makes it easy for them to be dismissed *in toto* – throwing the baby out with of the bathwater, you could say. However, if we examine the hard phenomena of ESP in separation, then it is much more difficult to dismiss them.

And it is these phenomena that I'm going to examine in this chapter. In fact, I'm mainly going to focus on two types of hard paranormal phenomena: telepathy and precognition. This is mainly for reasons of space. Psi is such a vast area that it would be impossible to cover every aspect of it and I've chosen these two areas in particular because they have both been investigated so rigorously, over many decades.

From Chapter 3 onwards, each chapter of this book has followed the same basic pattern. A different "puzzle" (for example, consciousness, NDEs or spiritual experiences) has been introduced and then the attempts to explain the phenomenon from a materialistic perspective have been examined. Weaknesses in these explanations have been highlighted, then an alternative panspiritist explanation has been suggested. All along, I've tried to show that the panspiritist explanation is less problematic and more elegant and logical than the materialist one.

But because psi is such a controversial subject, this chapter has a different structure. Materialism doesn't doubt the *existence* of the phenomena which we've been looking at over the last few chapters, such as consciousness, the placebo effect, hypnosis and near-death experiences. It just believes that they can all be explained in simple neurological or physiological terms. But materialism doesn't just doubt the existence of psychic phenomena, it *vehemently denies* them.

The main problem is that, unlike the other phenomena we're looked at, psi cannot be explained in materialist terms. As we have seen, it's easy to come up with possible materialist explanations for consciousness, near-death experiences and awakening experiences (no matter how flawed these explanations might be). But this isn't the case with telepathy or precognition. Materialism's only option is to deny their existence – which usually means explaining away positive findings as being the result of fraud, coincidence or poor research practices.

This is probably why psi is so controversial – and infuriates materialists much more than any other phenomenon. The other phenomena don't necessarily threaten the materialist worldview, because they can be explained. But psi does threaten the materialist worldview. One of the most common arguments used by sceptics is that psi contravenes the laws of physics. As we will see shortly, this is a false argument. But what *is* certain is that psi contravenes – and therefore undermines – the basic principles of materialism.

So my main focus in this chapter will not be to offer an alternative explanation, but to make a case for the existence of ESP – telepathy and precognition in particular. Then I will explain why the existence of these phenomena makes sense from a spiritual (or panspiritist) perspective.

Some examples of precognition

Surveys suggest that between one-quarter and one-third of people report having had at least one precognitive experience, most of which (around 75 per cent, according to the pioneering psi researcher, the late JB Rhine) occur in dreams.[2] One well-known example is of a man called John Godley, who dreamt the names of winning racehorses. One night in 1946, when he was a student at Oxford University, Godley dreamt he was reading a list of horse race winners in a newspaper and saw the names Bindal and Juladdin. The next morning he checked a newspaper and found that two horses with those names were running that day. With a group of friends, he decided to risk a bet; both horses won and the group won large sums of money. This happened to Godley several times over the following years. The fourth time he had such a dream he made a written statement of his predictions (again involving two horses), which was witnessed by a number of people, sealed in an envelope, stamped by a post office official and locked away until the day of the race. When this prediction came true Godley became famous around the world.

Frequently, precognitive dreams feature disaster premonitions. One of the biggest catastrophes in Britain's modern history occurred in 1966, in the Welsh mining village of Aberfan, when a massive coal slip came down a mountainside and engulfed houses and a school, resulting in the death of 144 people, including 128 children. One psychiatrist who worked with the villagers after the disaster, JC Barker, found 76 people who claimed to have had premonitions of the accident. Some of these were just feelings of intense anxiety – a feeling that something terrible was going to happen – but others were actual visions of the events in dreams. In all 36 people "saw" the accident in their dreams, and in some cases the visions were so vivid that they woke up in fright.

It's well known that premonitions caused some people to cancel bookings on the *Titanic*; more recently, the scientist and psychic investigator Rupert Sheldrake has found that many people had premonitions of the 9/11 terrorist attack in New York.[3] One of the most striking examples of these "disaster premonitions" is a man called David Mandell, a retired lecturer who, in the 1990s, began to have vivid dreams which he believed contained visions of the future. Because he was an artist, he started to paint scenes from the dreams. Then he would go to his local bank to have them photographed underneath an electronic clock that showed the date and the year. Eventually, he realized that he was seeing visions of future world disasters. For example, one painting showed a gas attack at an underground station, and in his notes accompanying the picture Mandell mentioned that it would take place in Tokyo, and he referred to a "secret group" that came from the "hills". This seems to relate to the attack in 1995 by members of the religious cult Aum Shinrikyo, who released sarin gas into underground trains in Tokyo, killing 12 people. The group lived in the hills outside Tokyo. There was also a picture of a Concorde aeroplane crashing at an airport, anticipating the crash in Paris in 2000. Mandell had included a French flag in the picture, indicating that he was aware of the location. But most strikingly of all, on 11 September 1996 he photographed a painting that showed two burning towers falling into one another and also included an outline of the head of the Statue of Liberty and the silhouette of an aircraft flying downwards.

Mandell was happy to be investigated by scientists – he passed a lie detector, and forensic tests of the negatives of his photographs showed no signs of tampering. Chris French, a well-known sceptical psychic investigator, devised a test to ascertain the accuracy of the paintings: 20 people were shown Mandell's paintings and given his own interpretations of them together with alternative interpretations, and they were asked to choose which interpretation best suited the paintings. In 31 out of 40 pictures, all 20 people chose Mandell's interpretation – a result which French himself said was statistically very significant.[4]

Apart from David Mandell's case, the earlier examples are anecdotal, and so unlikely to convince anyone who is sceptical about psi. Quite rightly, scientists demand hard evidence using tightly controlled experiments that satisfy the highest standards of scientific research. Fortunately, this evidence is available too.

Empirical evidence for psi

In 2011 the eminent psychologist Daryl Bem – at the present time, professor emeritus at Cornell University – published a paper called "Feeling the Future" in a prestigious academic journal, the *Journal of Personality and Social Psychology*. The paper described the results of nine experiments, involving more than 1,000 participants, eight of which showed significant evidence for precognition. Across a variety of different procedures, Bem found that his participants seemed to be able to "intuit" information before it appeared. In a simple example, they were shown a pair of curtains on a computer screen and asked to click on the curtain where they thought an image would be. At that point an image was randomly generated, and equally likely to appear behind either of the curtains. It was found that a significant number of the participants chose the correct curtain. And since no image was actually there at the time the participants chose, this was seen as evidence of presentiment. The editors of the journal quite correctly remarked that Bem's findings "turn our traditional understanding of causality on its head".[5]

However, prominent sceptics of psi phenomena were outraged and dismissed Bem's findings out of hand. The psychologist Ray Hyman described the results as "pure craziness... an embarrassment for the entire field".[6] The physicist Robert Park called it "a waste of time... it leads the public off into strange directions that will be unproductive".[7] The science journalist Jim Schnabel characterized these responses as an attempt to "suppress the findings of a scientific colleague because his findings threatened his reality".[8]

Bem encouraged other researchers to repeat his experiments, and many did so over the next few years. Perhaps even more significantly than the original experiments, a meta-analysis of 90 attempted replications of the experiments (involving 12,406 participants in 33 different laboratories) showed a highly significant positive result. According to Bem, this provided "decisive evidence" for his experimental hypothesis that human beings can sense future events.[9]

In fact, these findings would not be surprising to anyone who is aware of the history of psi research. As far back as the 1930s, the researchers JB and Louisa Rhine found that volunteers could guess, with a success rate three million times higher than chance, which cards were going to be taken at random from a pack.[10] The psi researchers Charles Honorton and Diane C Ferrari analysed the results of 309 "forced choice" precognition experiments published between 1935 and 1977, involving more than 50,000 participants and published in 113 scientific articles. They found a highly significant success rate (the odds against which were 10^{24} to one), which far outweighed any possible bias due to selective reporting.[11] A meta-analysis of more recent presentiment experiments (measuring human physiology before the presentation of startling images), between 1978 and 2010, found an even more significant positive result.[12]

Research with individuals who seem to have a special "gift" for precognition (or clairvoyance or telepathy, since sometimes the phenomena are difficult to distinguish) ability have also provided highly significant positive results. For example, a Dutch man named Gerard Croiset claimed to have psychic abilities, and was tested very intensively over 25 years by members of the Dutch Society for Psychical Research. A researcher devised a simple test that was repeated several hundred times. Croiset was taken to a public venue, such as a theatre or meeting house, where a chair would be selected at random. Croiset would then have to describe the person who would next sit on the chair on a particular date. The descriptions Croiset gave were very specific, including details such as hair and eye colour, age, clothes, any unusual marks and details about their personality, interests and personal lives. The descriptions were recorded and locked away, and at the lecture or performance in question, the recordings were played back, then the person who was sitting on the chair would stand up and comment on the accuracy of what he or she had heard. Croiset's descriptions were amazingly accurate. Over the years, as he became more well known, he was tested by many other investigators and the experimental conditions were made more difficult, but Croiset continued to be successful.

Research on telepathy has had some similarly striking results. One of the standard experiments used to test for telepathy is the so-called "Ganzfeld" procedure. This has been used since the 1970s. Typically, there is a "receiver" who is deprived of external stimuli, wearing headphones and halved ping-pong balls over their eyes. A "sender" tries to send an image to the receiver, who describes what (if anything) they are receiving. Afterwards, the receiver is shown four images and asked to pick the one that was "sent" during the experiment.

Obviously, the expected rate of success according to chance would be 25 per cent. But over thousands of trials, the success rate has been significantly higher than this. A meta-analysis of more than 3,000 Ganzfeld trials that took place from 1974 to 2004 had a combined "hit rate" of 32 per cent. A rate that is seven per cent higher than chance may not seem so impressive, but over such a large number of experiments this equates to many billions to one.[13]

Interestingly, Ganzfeld experiments with creative people have showed a significantly higher-than-normal rate of success. In 128 Ganzfeld sessions with artistically gifted students at the University of Edinburgh, a 47 per cent success rate was obtained, with odds of 140 million to one.[14] Similarly, in a session with 20 undergraduates from New York's Juilliard School of performing arts, the students achieved a hit rate of 50 per cent.[15] Another study primarily with musicians had a 41 per cent success rate.[16] One theory that may help to explain these findings is that psychic abilities are related to "labile" or soft self-boundaries which enable people to be open to influences and energies beyond the normal conscious mind. This would also allow people to open up to artistic inspiration – and potentially to spiritual experiences too.

Many other experiments with telepathy have had significant results. Experiments by the scientist Rupert Sheldrake have demonstrated the reality of "telephone telepathy" and also that dogs may apparently have a telepathic connection with their owners. In a long series of experiments with a dog called Jaytee, Sheldrake found that it would sit by the window for a significant proportion of the time that her owner was on her way home – 55 per cent of the time compared to just 4 per cent during the rest of her absence.[17] (We will discuss this experiment later, in relation to its attempted replication by the sceptical researcher Richard Wiseman.)

As sceptics of psi often point out, we shouldn't necessarily accept these results at face value. It is possible that the positive results of some experiments – particularly earlier ones – were partly due to factors such as flawed methodology, poor controls or even fraud. But in recent times the protocols of psi experiments have become extremely stringent, partly in response to the criticisms of sceptics. Contemporary psi experiments (such as Daryl Bem's) are actually more rigorously conducted than many experiments in other fields, because researchers are aware of the controversial nature of their experiments and the intense scrutiny they will attract.

The "file drawer effect"

The "file drawer effect" is sometimes cited as a possible reason why meta-analyses of ESP experiments show significant positive results. What this refers to is that if an experiment comes up with negative results, it is less likely to be published. In the first place, researchers may be less likely to submit null studies for publication; even if they do, academic journals are less likely to publish them than they would positive studies. So, according to this argument, there are many ESP experiments with negative results which have disappeared into oblivion and therefore aren't included in the meta-analyses.

However, there are some valid reasons why the file drawer effect can't account for the positive results of ESP meta-analyses. Firstly, even early psi researchers were aware of the file drawer effect as a potential problem, and took measures to include it as a factor in analyses. JB Rhine devised a statistical method to calculate the effect of publication bias, and in 1975 the Parapsychological Association adopted a policy of publishing null results so that they would always be included in meta-analyses. Following on from these measures, it has become common practice to measure study selection bias in psi research, using a variety of techniques.

And, significantly, even when such techniques are applied meta-analyses still show significant positive results. In the meta-analysis of more than 3,000 trials from 1974 to 2004 that I mentioned earlier, researchers calculated that 2,002 unpublished null studies would have been required to reduce them to insignificance. The average Ganzfeld study over this 30-year period consisted of 36 trials, and so the 2,002 other studies would have required 72,072 additional sessions – which would mean running Ganzfeld sessions continually 24 hours a day for 36 years.[18] In an analysis of 27 more recent Ganzfeld studies, researchers calculated that it would take 384 unpublished papers to nullify the results. Here it's important to consider that, in recent times (partly due to problems obtaining funding) only a fairly small number of psi experiments have been carried out. Because the Parapsychological Association has had a policy of publishing null studies, it seems unlikely that such a large number of studies could have been performed then secretly filed away.[19]

Replication

Another issue that is often raised about positive psi findings is what scientists call replication. This is when experiments are repeated to find out if their results are reliable. A researcher could get a positive result in an experiment, only for other researchers to get different results, which would suggest that the original experiment was flawed in some way.

In practice, psi experiments have done pretty well at replication. Many experiments – such as the Ganzfeld tests, and Daryl Bem's experiments – have been repeatedly replicated with success. But some materialistic scientists set the bar very high when it comes to psi, claiming that psi phenomena can't be considered proved because they have not been replicated *without fail*. Just one experimental failure is seen as justification for invalidating a whole series of positive results. For example, a team of sceptical psychologists led by the UK researcher Stuart Ritchie repeated one of Daryl Bem's experiments, but didn't have positive results. And despite the many other successful replications, this one failure was seen as justification for rejecting Bem's "feeling the future" hypothesis outright.[20]

However, in other areas of science, it's very rare for a "one strike and you're out" policy to be applied. It's not uncommon for research to be tacitly accepted before it has been replicated. In some cases, replication is never even attempted. And when it is attempted the rates of successful replications are typically extremely low. One team of researchers found a 25 per cent replication rate,[21] while others found an even lower rate of 11 per cent.[22] Compared to this, the replication rates of many psi experiments are impressive – as the earlier meta-analyses indicate. The thousands of repeated Ganzfeld and autoganzfeld (which is when the procedure is controlled by a computer) experiments since the 1970s with above-chance results surely do constitute successful replication.

One argument that could be applied here is that because psi is so controversial, the bar has to be set high – as in the phrase (usually credited to the astronomer Carl Sagan) "extraordinary claims require extraordinary evidence". However, this seems prejudicial, like saying that certain types of criminals need higher levels of evidence to prove their innocence. It is also debatable whether psi actually *is* extraordinary. Research has shown that most people believe they have experienced at least one form of psi, and most cultures throughout history have accepted it as a reality. It only seems extraordinary in the context of materialism.

It's also important to remember that ESP is not, by its nature, completely constant or reliable. You can't compare testing for telepathy or precognition to testing "standard" psychological processes such as attention, perception or memory. Psychic abilities vary from person to person. In some people they don't appear to exist at all, whereas others (such as creative people, as we have seen) may possess them to a high degree. Psychic abilities may also be situational; even with a person who normally demonstrates them to a high degree, there may be some circumstances when they fail – for example, when they are nervous or stressed.

In this sense, you could compare ESP abilities to creative abilities like painting or writing poetry. Some people have very little ability in these areas, perhaps none at all. Some people might be able to do them passably, and some people – probably the smallest group – are very skilled in them. And whether people do demonstrate their creative abilities is situational; even a very skilled creative person may not be able to demonstrate his or her creativity in an uncongenial environment, in which they feel uneasy. Both ESP and creative abilities work best in states of calm and relaxation. (And as we saw earlier, ESP and creative abilities may be connected in that they are both linked to "labile" self-boundaries.)

As a result, it's not surprising that replications of ESP experiments are not always successful. To expect otherwise would be like expecting all human beings to reliably demonstrate poetic abilities in laboratory experiments. And it is very impressive that, despite this constraint, psi experiments *do* have relatively high rates of replication.

Beyond our normal awareness

One reason why I personally have always been open to the possibility of psi is that I don't believe that we can, in William

James's phrase, close our "accounts with reality". As I have already pointed out, some materialist scientists tacitly assume that our present vision of reality is fairly reliable and complete. While leaving some room open for new additions and amendments to our present understanding, they believe that the main elements of our worldview are solidly established. They assume that we have discovered the major forces and natural laws by which the world operates.

But I believe this is foolish and arrogant. Every animal has a limited awareness of reality. For example, think of our awareness of reality compared to a sheep's. We are aware of many phenomena and concepts which a sheep is probably not aware of; for example, the concepts of death, the future and the past. But although we may have a more intense awareness of reality than most other animals, it is extremely unlikely that our awareness is complete. To believe otherwise is a form of anthropocentrism, tantamount to seeing human beings as the end point of the evolutionary process. It is possible that at some point in the future other living beings will come into existence who will have a more intense awareness than us, just as we have more intense awareness than sheep. These hypothetical beings may be more intensely aware than we are of the phenomenal world around them, and *more* aware of phenomena – for example, forces, energies or laws – which are "selected out" of our awareness or beyond its range.

So it is probable that there are, in Shakespeare's phrase, "more things in heaven and earth… than are dreamt of in [our] philosophy". There are almost certainly forces, energies and other phenomena in the universe beyond those which we can presently perceive and understand, or even detect. We may be aware of the effects of some of these phenomena without being aware of the phenomena themselves – in the same way that, for example, an insect may be able to sense the heat of the sun without being aware of the sun as an entity in itself. This is clearly indicated by quantum physics. There are so many well-established aspects of quantum physics (such as entanglement, "quantum tunnelling" and wave/particle duality – all of which we will look at shortly) that seem to make no sense in terms of our normal understanding of the world. We know that they happen but we don't understand *how* they happen – that is, we don't understand the underlying forces or principles that allow them to happen. And this also applies to the forces or principles that generate – or would explain – telepathy and precognition. Perhaps one day we will understand these forces; or perhaps they will remain forever beyond our awareness and our comprehension. In any case, the dogmatic assertion that psi doesn't – and cannot – exist is probably a good example of what the philosopher Alfred North Whitehead described as: "the self-satisfied dogmatism with which mankind at each period of its history cherishes the delusion of the finality of its existing modes of knowledge".[23]

Does psi break the laws of science?

The above discussion links to a couple of other arguments that are often used against psi. The first argument is that it's invalid because we don't have a theory to explain how it could work. However, there are many things that scientists accept as real without understanding how they work. The most obvious example is consciousness. Daniel Dennett and a few other theorists may argue that consciousness is an illusion, but the majority of theorists who study it agree that they are investigating a real phenomenon, at the same time as admitting that we don't yet have an adequate theory to explain it. The same is true of some of the quantum phenomena I've just mentioned. So it is illogical to "outlaw" a phenomenon just because we can't explain it.

The second argument is that psi can't exist because it contravenes the fundamental principles of science. If it really existed, so it is argued, we would have to completely revise our understanding of physics. The sceptic Douglas Hofstadter argued along these lines in response to Daryl Bem's work, writing that if its results were valid, it "would necessarily send all of science as we know it crashing to the ground".[24]

However, although psi might contravene some of the laws of traditional Newtonian physics, it is completely compatible with many of the findings of modern physics.

For example, take precognition. The idea that the future may already exist in some way makes no sense in Newtonian terms. but it is completely compatible with more recent concepts in physics. Many modern physicists do not think of time in a linear way, but speak of the *spatialization* of time, or of a "block universe" in which everything that has ever happened or ever will happen exists simultaneously. The universe exists in four-dimensional space-time, and in these terms time, rather than flowing, is just there in the same static way that space is. In four-dimensional space-time, the concepts of past, present and future are meaningless. According to the physicist Roger Penrose: "The way in which time is treated in physics is not essentially different from the way in which space is treated... We just have a static-looking fixed 'space-time' in which the events of our universe are laid out!"[25] Penrose goes on to suggest that our linear sense of time passing forwards is illusory, created by our minds in order to impose order on our experience. (This is similar to the philosopher Immanuel Kant's suggestion that time and space are not innate aspects of the world, but "categories" created by the human mind.) The idea that time passes, or that there is a past and a future divided by a present, has never been verified by any physical experiments. As Paul Davies, another physicist, noted: "As soon as the objective world of reality is considered, the passage of time disappears like a ghost in the night."[26]

This is also implied by Einstein's theory of relativity. As well as showing that time is relative to speed and gravity, Einstein showed that there is no set order in which events occur. From one person's vantage point two events might seem to happen at the same time, but to someone moving at high speed there might be a time lag between them. Or there might be two events that happen one after the other, which one person who is travelling at a high speed sees in reverse order to another. Again, this suggests that, in Einstein's own words, "the distinction between past, present and future is only an illusion, even if a stubborn one."[27]

This is also strongly suggested by some of the findings of quantum physics. In the microcosmic sub-atomic world, time seems to be indeterminate. Some quantum physicists claim that at this level of reality, cause and effect can sometimes be reversed, which means that an event can literally take place *before* its cause. There is even a term for this: retrocausation. As the physicist Daniel Sheehan notes: "It seems untenable to assert that time-reverse causation (retrocausation) cannot occur, even though it temporarily runs counter to the macroscopic arrow of time."[28] This notion was also accepted by the founders of quantum physics such as Pascual Jordan and Werner Heisenberg. After noting how the normal sequence of events is sometimes reversed in the quantum world, Jordan remarked: "This has enormous implications for psychology and parapsychology, since such reversals of the cause-and-effect sequence are proved possible and philosophically valid."[29] Similarly, the "transactional" interpretation of quantum mechanics sees quantum events as an interaction between waves that move both forwards and backwards in time. The originator of this approach, American physicist John Cramer, refers to a quantum event as a "handshake" across space-time, where a wave is "offered" out and it is "confirmed" by another wave from the future. This means that, at the quantum level, time is a two-way street, in which the future determines the past, as well as vice-versa.[30]

It is also very debatable whether telepathy breaks the laws of physics. The quantum phenomenon of entanglement means that once two particles have interacted they will always remain connected. No matter how far apart, they will spin together in harmony and balance each other's random fluctuations. In other words, particles can't be treated as separate entities but only as a part of a whole system. This suggests a fundamental level of interconnection that would allow for the possibility of an exchange of information via telepathy.

Psychokinesis – when physical systems are influenced by mental intention, such as when a person uses mental powers to alter the movement of objects, like influencing the fall of coins or stopping a watch hand – is also conceivable in terms of quantum physics, albeit in a less obvious way. In the sub-atomic world, there is no separation between our own consciousness and the world "out there". The traditional idea of the "objective" scientist who stands apart from his experiments, observing and recording results with complete detachment, is revealed to be a myth. In the quantum world, the observer always interacts with what he or she is observing, changing the behaviour of particles through the acts of observation and measurement. As Pascual Jordan put it, "Observations not only disturb what has to be measured, they produce it."[31] And this notion of the mind interacting with and influencing particles isn't so far away from psychokinesis. Quantum physics shows that consciousness and the "material" world are intertwined.

It is also very important to note that the strangeness of quantum physics isn't just confined to the microcosmic sub-atomic world. The relatively new field of quantum biology suggests that large-scale quantum effects happen everywhere around (and inside) us. Biological processes such as photosynthesis, olfaction (the sense of smell) and the reaction rates of enzymes may take place via entanglement and "quantum tunnelling" (where particles pass through solid barriers in a way that is impossible according to Newtonian physics).[32] These phenomena blur the distinction between the "supernatural" and the "natural", making as little sense in terms of classical physics – and materialism in general – as telepathy or precognition. (We will look at quantum biology – and quantum physics in general – in more detail in Chapter 11.)

It is therefore invalid to say that telepathy and precognition contravene the laws of physics, or would require a new model of science. In fact, it could even be argued that in modern physics, psi isn't just conceivable but *inevitable*. As the physicist Olivier Costa de Beauregard wrote: "relativistic quantum mechanics is a conceptual scheme where phenomena such as psychokinesis or telepathy, far from being irrational, should, on the contrary, be expected as very rational."[33]

Significantly, the argument that psi contravenes the laws of psychics is usually put forward by psychologists and philosophers rather than physicists themselves. In fact, many physicists (such as Marie Curie, Wolfgang Pauli, Max Planck, Eugene Wigner, JJ Thomson, David Bohm and John Stewart Bell) have been open to the possibility of psi.

Psi and panspiritism

As with modern physics, in panspiritist terms psi makes a great deal of sense. In its essence, spirit-force is timeless. As Kant pointed out, time is a mental construct. It only comes into existence with our individuated consciousness. This is something we touched on in the previous chapter, in relation to awakening experiences. At the highest intensity of awakening experiences (and also in near-death experiences) time ceases to exist, because we touch into the essential timelessness of spirit-force. In the same way that spirit-force is everywhere in space, it is everywhere *in time*. So, in a sense, the future has already happened – and, indeed, the past is still happening. The past doesn't pass away into nothingness, and the future isn't waiting unknowably in front of us. All of it is here and now, just as it is in the concept of the "block universe" in physics. So precognitive experiences, like high-intensity awakening experiences, are simply a transcendence of the illusion of linear time. (This is also true of the phenomenon of retrocognition, when the past is experienced directly in the present.)

Telepathy clearly makes sense in terms of panspiritism too. Again, there is a parallel with awakening experiences. In higher-intensity awakening experiences we transcend separateness and individuality in the same way that we transcend time. We become one with other human beings (sometimes other living beings too), as well as with nature, and possibly the whole cosmos. In awakening experiences we experience this oneness in terms of being. But in telepathy we experience it in an *informational* (or cognitive) sense. From the panspiritist perspective there is no separation between human minds. So it is not surprising that occasionally we are able to enter into each other's mental space and sense each other's thoughts and intentions.

Psychokinesis also makes sense from a panspiritist perspective. Spirit pervades everything, and it connects us to everything – not just to other human beings, but to all things. So in a sense we *are* everything (as again is implied by the "observer effect" in quantum physics). In theory, then, we should be able to influence the structure of physical things. We saw in Chapter 5 that our mental intentions frequently influence our bodies, and psychokinesis can be seen as an extension of this. The puzzling issue is perhaps that psychokinesis happens so seldom compared to mind–body influences. But as I have already suggested (in Chapter 5), the important difference may be that we are much more intimately associated with the matter of our bodies than with matter in general. Our own minds are so closely intermingled with the matter of our body that it's impossible to separate them. We shouldn't really speak of a mind and body, but only a mind/body complex. So it's not surprising that our own thoughts and beliefs can influence the functioning of our body so strongly. But when it comes to physical things that are not intimately related to us, the effect is inevitably much less common.

Acceptance and denial

When I tell people that there is very robust evidence for psi, they sometimes ask me, "So why isn't it widely known? Why is it rejected by scientists?"

First of all, it's important to point out that psi isn't rejected by all scientists. The evidence for psi – together with the theoretical possibility of its existence – has convinced some more open-minded scientists, who weren't so in thrall to the paradigm of materialism. In 1949 the mathematical genius Alan Turing wrote of psi: "These disturbing phenomena seem to deny all our usual scientific ideas. How we should like to discredit them! Unfortunately the statistical evidence, at least for telepathy, is overwhelming. It is very difficult to rearrange one's ideas so as to fit these new facts in."[34] Even Einstein (who was less convinced about psi than other physicists) wrote: "we have no right to rule out a priori the possibility of telepathy. For that the foundations of our science are too uncertain and incomplete."[35]

In fact, in their more unguarded moments even the most ardent materialists (and sceptics of psi) have admitted that psi experiments have highly significant results. The British psychologist Richard Wiseman – one of the most active and vociferous critics of psi – has remarked that "by the standards of any other area of science remote viewing is proven."[36] In 1996 the statistician Jessica Utts (not herself a sceptic) stated: "using the standards applied to any other area of science, it is concluded that psychic functioning has been well-established."[37] Surprisingly, one of America's most ardent critics of psi research, the philosopher Ray Hyman (who reacted with outrage to Daryl Bem's findings), agreed with Utts, writing that the research findings "do seem to indicate that something beyond odd statistical hiccups is taking place. I also have to admit that I do not have ready explanation for these observed effects."[38]

These moments of relative open-mindedness contrast with the attempts of some materialists to deny – or explain away – the positive results of their own psi experiments. In 2005 researchers at Notre Dame University conducted a series of eight Ganzfeld experiments, which found a highly significant overall "hit rate" of 32 per cent. The researchers admitted that, as sceptics, this result made them feel "uncomfortable", because it came "precariously close to demonstrating that humans do have psychic powers".[39] In reaction they developed a further experiment, where they matched up individuals who had "hits" during the previous eight experiments. For some strange reason, these pairs produced the highly significant *negative* result of a 13 per cent hit rate (significantly lower than the 25 per cent chance rate). Encouraged by this negative result, the researchers claimed that it invalidated the previous eight experiments and concluded that they had found no evidence for the existence of telepathy.

Similarly, there was a great deal of controversy when Richard Wiseman attempted to replicate one of Rupert Sheldrake's experiments with the dog Jaytee (as described earlier). According to the methodology used by Sheldrake, Wiseman's four experiments actually yielded a *more* positive result than Sheldrake's – Jaytee sat by the window 78 per cent of the time that her owner was travelling home, compared to 4 per cent during the rest of her absence. (In Sheldrake's experiments, it was 55 per cent, compared to 4 per cent during the rest of the owner's absence.)[40]

That would seem to be an incontrovertible successful replication of Sheldrake's experiments. However, Wiseman chose to ignore this data and used a different criterion of success: Jaytee had to go to sit by the window *at the exact moment* that her owner set off home. If Jaytee went to the window before this, this would mean that she had "failed". And, not surprisingly, by this criterion the experiments were judged to be unsuccessful and bizarrely presented as "proof" that Jaytee (and dogs in general) do not have psychic powers.[41]

This shows that it's always possible to explain away evidence if you don't like it. If you are strongly attached to a belief system, any evidence that seems to contradict it creates cognitive dissonance, which in turn generates an impulse to "bury" that evidence. This is not dissimilar to the way that creationists try to explain the existence of fossils by saying that they were put there by God to test our faith (or by Satan to tempt us into unbelief).

This takes us back to the point I made in Chapter 1, about the important psychological function of belief systems: they provide a narrative to make sense of our lives, with a sense of orientation and understanding of our predicament. At the same time, to feel that we can explain the world provides a satisfying feeling of control, a sense that nature is "under our thumb" and in thrall to us. As a result, we are very reluctant to accept evidence that contravenes our belief systems.

The enlightenment project

Some materialists see themselves as a part of a historical "enlightenment project" whose aim is to overcome superstition and irrationality. The Enlightenment movement that first flourished during the 18th century was originally a process of liberation from the hegemony of the Church and monarchy, replacing dogma and myth with scientific knowledge. And there is no doubt that this project has been massively beneficial to the human race – giving rise to medicine, technology, freedom from superstitions and taboos, and from social and intellectual oppression, resulting in a truer, more evidence-based concept of reality.

However, for some of its adherents, the enlightenment project leads to a blanket opposition to any phenomena which appear to be "irrational". More specifically, it has led to a "category error" of associating phenomena such as telepathy and precognition (or indeed, spiritual experiences or near-death experiences) with phenomena such as fundamentalist Christianity or Islam, or superstitions and taboos. This is an example of what the philosopher Ken Wilber calls the "pre/trans fallacy", which explains that it is very easy to confuse "pre-rational" and "trans-rational" phenomena.[42] (Wilber gives the example of Freud's attitude to spiritual experiences, which he explained in terms of a regression to an infantile state of oneness with the mother. In other words, Freud explained a "trans-rational" experience in "pre-rational" terms.) Fundamentalist religion can be seen as a "pre-rational" phenomenon, since it wilfully ignores the evidence of science (for example, with respect to evolution) and clings to a mythic view of reality. But phenomena such as telepathy and precognition – for which there is a lot of empirical evidence and which do accord with the findings of quantum physics and theories of consciousness – are better seen as "trans-rational". That is, they aren't related to ignorance or superstition, but to unknown phenomena or forces which are – at least at present – beyond the limits of our awareness. They are not beneath us, but beyond us.

Phenomena such as telepathy and precognition do not at all contravene the principles of science itself – only the principles of materialism. And since we have seen throughout this book that the principles of materialism are invalid, this is not problematic at all. Psi phenomena are real – their existence makes sense from a theoretical point of view and even from a scientific point of view. Moreover, there is ample experimental evidence for their existence. It is only the materialist worldview that deems psi to be "supernatural" or "paranormal", when it should really be seen as natural and normal.

CHAPTER 9

COMPLEXITY AND CONSCIOUSNESS: PUZZLES OF EVOLUTION

In 1952 a young graduate student called Stanley Miller discovered that amino acids – the basic building blocks of life – could be spontaneously synthesized in a chemical simulation of the Earth's atmosphere. After this, many scientists believed that the problem of the "origin of life" would soon be solved. It seemed inevitable that similar experiments would lead to the formation of a "self-replicating" molecule from those amino acids.

However, soon after Miller's experiment, researchers began to realize that life was actually a lot more complicated than they had presumed. Cells were not simply little packets of chemicals, but highly complex and intricately organized entities. Another problem identified by researchers was the "chicken and egg" problem that DNA needs proteins to function, but protein itself needs DNA to be sequenced properly. So how could one arise without the other?

Some researchers noted the problem that, because life is so complex, and relies on the combination of so many different elements in such an intricate way (all of them building on each other in layers of increasing organization and interconnectivity), there may not have been enough time for it to emerge accidentally on the Earth. Such theorists pointed out that there was only a relatively short "window of opportunity" for life to emerge. The Earth came into being around 4.5 billion years ago, and we know that until 3.9 billion years ago it was heavily bombarded with meteors, which would have made life impossible. And yet the first signs of life appear in fossil records 3.8 billion years ago. This means that life had no more than 100 million years to come into being. This led to the theory – one of whose adherents was Francis Crick, whose neurological theories of consciousness we discussed in Chapter 3 – of "panspermia", which suggests that life may actually have begun elsewhere in the universe, and that the Earth may have been "fertilized" from interstellar space. However, as other scientists have pointed out, the odds against this may be even greater than the odds against life starting accidentally on this planet.

In recent years, the most popular theory has been that life originated with RNA molecules. (Like DNA, RNA is a nucleic acid that has a role in decoding and expressing genes.) The standard way of thinking about the origin of life was of dozens of molecules coming together simultaneously in a chemical soup. But perhaps life didn't depend on all of these different molecules. RNA molecules have been found to be very versatile, and could perhaps have performed all of the functions of life on their own. In 2009 this theory was given a big boost when scientists managed to synthesize RNA enzymes that could replicate themselves.

However, some scientists remain sceptical about this theory, mainly because it seems impossible that a complex and sensitive molecule such as RNA could have come into existence spontaneously in the volatile environment of the "prebiotic" world (that is, the world before life began). Another problem is that in order to act as catalysts (enabling chemical reactions with other molecules), RNA molecules would have to take the form of long, folded chains – and, again, the prebiotic environment would have prevented this.[1] Other scientists have argued that RNA can't be the source of life because it lacks the "reflexivity" – the ability to form a feedback loop – that is needed for life forms to emerge.[2]

As a result, there is still no consensus about how life might have formed accidentally. A dizzying range of different theories have been put forward, most of which are highly speculative. In 2017 alone at least three different theories were proposed: one that life originated as a partnership between RNA and peptides;[3] another that life must have begun with proteins, which are much more likely to emerge spontaneously than nucleic acids such as RNA;[4] and, finally, a theory that phosphates were essential to the origins of life.[5]

In other words, more than six decades after Miller's original experiment, there is no consensus on – and no satisfactory explanation for – how life might have formed. Despite this, the idea that life on Earth came into being by accident is still taken for granted by most scientists. This perhaps isn't surprising, since the alternative to this seems to be creationism. If life didn't start accidentally, then wouldn't we have to believe that God created living beings?

Evolution

Once the first self-replicating molecule emerged – by whatever means – life began a slow process of evolving into different forms. Evolution is usually seen as a process by which, over vast periods of time, living beings diverge into more and more varieties, and become increasingly complex and more intricately organized. Over hundreds of millions of years, single-celled bacteria evolved into more complex multi-celled creatures, which led to life forms such as sponges, fungi, corals and sea anemones, and then to insects, fish and land plants.

As evolution progressed there was an increasing specialization and differentiation – more and more cells collected together and worked together more intricately, with different roles. This led to the development of the first brains (in flatworms). Over time, brains grew larger and larger, and their cells became more interconnected. And after many millennia of gradually increasing complexity, the process led to the emergence of highly complex mammals with tens of billions of brain cells, such as human beings.

According to the standard materialist view, evolution is an accidental process that happens due to a combination of two main factors: mutations and natural selection. When life forms reproduce, genes are copied into offspring. Occasionally, a gene isn't copied correctly and a mutation occurs. In almost all cases mutations have negative effects, and the life forms who have them don't reproduce, and so quickly die out. But very occasionally, mutations have a beneficial effect, giving life forms an advantage in survival. These positive mutations are "selected" and become established in the gene pool. This is what creates variations in species, and over long periods of time such variations build up into major differences, and eventually into different species.

This seems like a simple and logical theory, but in this chapter we will see that there are some problematic aspects to it – in particular, the idea that the process has been completely random and accidental.

The heretical philosopher

In 2012 one of America's leading philosophers, Thomas Nagel, argued against the orthodox view of evolution in a book boldly titled *Mind and Cosmos: Why the Materialist Neo-Darwinian Conception of Nature Is Almost Certainly False.* Nagel described his scepticism about the assumption that life began accidentally, arguing that it was highly unlikely that "self-reproducing life forms should have come into existence spontaneously".[6] He was also dismissive of the idea that, once it had come into being, life's evolution into more and more complex forms could be explained in terms of mutations and natural selection. He found it implausible that "life as we know it is the result of a sequence of physical accidents together with the mechanism of natural selection".[7]

Nagel suggested that Neo-Darwinism (as the materialist view of evolution is called) has become a kind of belief system, little more than "a schema for explanation, supported by some examples".[8] He also pointed out that, as I stated in the introduction, many intellectuals feel obliged to accept materialism because they see it – falsely – as the only alternative to religion. As a result, they have learned to ignore the implausibility of many materialist assumptions. Nagel proposed an alternative to materialism, a worldview in which, "Mind is not just an afterthought or an accident or add on, but a basic aspect of nature."[9] In other words, Nagel was advocating a post-materialist view of the world, and a form of panpsychism, or perhaps even panspiritism. He also suggested that the world has a natural tendency to move towards greater value and complexity, and so the origins of life and evolution were not accidental, but inevitable and purposeful.

Many of Nagel's academic peers reacted with dismay to his views, and publicly disparaged him. The well-known cognitive psychologist – and adherent of evolutionary psychology – Steven Pinker tweeted disdainfully, "What has gotten into Thomas Nagel?", and spoke of "the shoddy reasoning of a once-great thinker".[10] The philosopher Simon Blackburn remarked tartly that, "if there were a philosophical Vatican, the book would be a good candidate for going on to the Index."[11] Other critics bemoaned that Nagel's book would give ammunition and encouragement to the fundamentalist religious opponents of Darwinism.

These comments show the quasi-religious fervour with which materialists sometimes react to the questioning of their tenets, which is an indication that they are defending an ideology rather than objective scientific findings. Simon Blackburn even uses a religious analogy, unconsciously revealing the ideological nature of his position.

In this chapter we will explore the problematic aspects that have made thinkers like Nagel doubtful about the Neo-Darwinist view of evolution.

Are mutations and natural selection enough?

Towards the end of his life, Charles Darwin came to regret that he had placed so much emphasis on natural selection in his theory of evolution. Although he still believed that natural selection was the main way in which variety had arisen in evolution, he harboured serious doubts that it was the *only* way. He didn't believe that natural selection was sufficient to account for the variety of life forms on Earth, and the seeming ease with which they arise.

Darwin's doubts about natural selection have never been fully resolved. Numerous observers have pointed out that it seems implausible that such a staggeringly positive and creative process could be generated purely by a negative phenomenon such as natural selection. It has been estimated that mutations only occur at a rate of about one per several million cells in every generation. However, since only a tiny number create beneficial traits which give a survival advantage, some scientists have suggested that this frequency is insufficient to account for the evolution of such a massive diversity of life forms.[12]

This isn't just because mutations happen so rarely, but also because in order to create significant changes (including the generation of new species) long series of beneficial mutations have to occur *in sequence*. Mutations have to be cumulative, perfectly matched to previous mutations, and occurring at the right place and time. So with every "matched" mutation, the odds against its random occurrence increase massively. In the words of the eminent French zoologist Pierre-Paul Grassé, mutations only "occur incoherently. They are not complementary to one another, nor are they cumulative in successive generations towards a given direction."[13]

Another problematic issue is explaining how natural selection can give rise to new structures and features, and especially new species. The standard view is that random mutations slowly create more and more variety over millions of years, and eventually these differences build up into distinct, new species. But it may not be quite as simple as this. As Grassé also pointed out, mutations only cause trivial changes. They are equivalent to "a typing error made in copying a text" with very little "constructive capacity" or innovation, so that they cannot create complex organs or body parts.[14] There are invisible boundaries between species which mutations cannot cross, meaning that they can cause variation but never true evolution.

There is also the problem that favourable mutations would soon be lost by interbreeding with non-mutated members of a species. Darwin himself saw this as the biggest problem of his theory, and Neo-Darwinists have never convincingly solved it. It's easy to see how this "crossing" might be avoided with animals, which could simply move away from one another, but not with the vegetable kingdom.

A number of biologists – such as Stuart Kauffman and Robert Reid – have argued that natural selection isn't sufficient to explain the arising of genetic variation and new life forms, and have proposed an alternative concept of "emergent evolution".[15] As with Thomas Nagel's view, this concept suggests that biological systems have an innate tendency to move towards greater value and complexity. (We will examine this concept of emergence in more detail shortly.)

The concept of "punctuated equilibrium" also casts doubt on Neo-Darwinism. Fossil evidence shows that evolution works through stops and starts, with periods of stasis for millions of years and then sudden bursts of change – which can be as short as 1,000 years – which give rise to new species. This doesn't fit well with the idea of random mutations, since these would surely occur fairly evenly. There would be no reason why some periods would see more change than others.

An alternative view of the origins of life

So what's the alternative? As I have suggested throughout this book, rejecting the standard materialist explanation doesn't mean we have to accept a religious interpretation. I am certainly not doubting the fact that evolution has occurred and advocating creationism. Right at the start of this book

I pointed out that just because I'm critical of materialistic science doesn't mean that I'm against science itself. The same is true of my attitude to evolution. I find it wonderful and miraculous that all life forms alive on the surface of the planet now are descended from simple, single-celled life forms that emerged around four billion years ago. The issue isn't about whether evolution has happened, but *how* it has happened.

Let me also say that I'm not going to suggest that panspiritism can completely – or even significantly – explain the origins of life or evolution. I'm going to suggest tentative explanations, which still leave a lot unaccounted for. I admit that my account is more of a descriptive than an explanatory one. As a result, I'm aware that the explanations I put forward in this chapter will be some of the most speculative of the book.

This is particularly the case with the origins of life. All we can really say from a panspiritist point of view is that the origin of life was connected to *the increasing complexity of physical forms*. Life began when, at a certain point, physical forms became complex enough to receive and transmit consciousness. As mentioned in Chapter 2, there is a distinction between simple physical structures that are pervaded with consciousness and more complex physical structures that can receive and transmit consciousness into themselves, and so become individually conscious. In this sense, all physical things have consciousness in them, but only more complex things actually *are* conscious. So when – with the formation of the first simple prokaryote cells that were protected from their own environment and had self-contained biochemical activity – a certain threshold of complexity was reached, universal consciousness was able to express itself in individual structures so that they attained a rudimentary awareness, and inner life.

Of course, this doesn't explain the mechanisms by which cells were (and are) able to receive and canalize consciousness, or how the cells emerged in the first place. But perhaps a discussion of evolution from the panspiritist point of view may be able to shed some light here – on the second problem, at least.

An alternative view of evolution

From the panspiritist point of view, evolution is not a random and accidental process but one that has an *impetus* behind it, a tendency to move towards increased complexity and increased awareness. As Thomas Nagel suggests, evolution is a teleological process – that is, it moves in a certain direction, with a certain purpose. And this teleological nature is driven by the innate tendency of consciousness to expand and intensify itself. Once universal consciousness was canalized into material structures, making them alive, it impelled those structures to become more complex and highly organized so that they could support more advanced forms of mentation, greater degrees of sentience, and more intensified and expansive forms of awareness.

Perhaps this tendency towards complexity existed – and expressed itself – even *before* the first life forms emerged. As a dynamic force that pervades all material things, spirit-force has always – right from the beginning of the universe – impelled life forms to become more organized and to develop into more complex forms. Perhaps this tendency of matter to move towards greater complexity was involved in the formation of the solar system, and the Earth, and then in the origin of life. In other words, it may be that the mysterious emergence of self-replicating cells out of simpler biological molecules was not an accident, but impelled by this tendency towards complexity. (This is essentially the same as Thomas Nagel's argument, as described earlier.)

In terms of biological evolution, the first important point to consider is that evolution has an *inner* dimension. Neo-Darwinists usually see evolution only in its outward, physical expression. But evolution doesn't just bring increasingly physical complexity, it also brings increased awareness. As living beings become more physically complex and more highly organized, they also develop more inwardness, a greater degree of inner life. They become more sentient, with a more intense awareness.

In this sense, at the same time as being one of the most physically complex species on this planet, human beings are probably one of the most intensely aware and sentient species. (I'm being careful to say "one of" because some species of whales and dolphins have more brain cells than us, and may be at least as conscious as we are.) It certainly appears that we have a more intricate and expansive awareness of reality than most other animals.

Here it's useful to think in terms of four different types of awareness: perceptual, intersubjective, subjective and conceptual. (Intersubjective simply means being connected to – and aware of – other beings.) All life forms appear to have a degree of perceptual awareness, in that they are aware of and respond to changes in their environment, such as when an amoeba moves towards sources of food, or to light. Many animals also have a degree of intersubjective awareness, through which they are connected to other members of their groups (a good example is the "swarm behaviour" of ant colonies, termites, flocks of birds and schools of fish). Some animals seem to have a degree of subjective awareness (or self-awareness) too. For example, as we saw in Chapter 3, a number of species – mostly primates, but Eurasian magpies and elephants too – have shown the ability to recognize themselves in mirrors, and to be aware of their own bodies. Some primate species have shown a degree of conceptual awareness too, in that they can be taught a rudimentary awareness of categories and numbers.

However, it seems reasonable to say that human beings possess these forms of awareness to a very intense degree compared to most other species. The amazing intricacy and complexity of human language, compared to the apparently rudimentary languages of some animals, testifies to the richness of our conceptual and subjective awareness, with our ability to deeply examine our inner subjective world and our intricate understanding of the world we inhabit. And whereas the intersubjectivity of some animals only appears to extend as far as a particular group (in swarm behaviour), human intersubjectivity is wide ranging and indiscriminate, even extending to other species (such as when identification and empathy with animals leads to vegetarianism or veganism). Human intersubjectivity may also be deeper, involving a powerful sense of compassion.

So, since the beginnings of life 3.8 billion years ago, the awareness of living beings has been intensifying and expanding across these four categories, in parallel with their increasing physical complexity. The increasing physical complexity has supported the development of increasing awareness, sentience and autonomy.

Other views of spiritual evolution

I'm certainly not the first person to put forward a spiritual view of evolution. Many philosophers have suggested that evolution is a purposeful process of the unfolding and intensification of consciousness, including the German philosophers Georg Hegel and Johann Gottlieb Fichte, the French philosophers Henri Bergson and Pierre Teilhard de Chardin, the Indian philosopher Sri Aurobindo and the contemporary American philosophers Ken Wilber and Michael Murphy. Teilhard de Chardin saw evolution as a process of the "spiritualization" of matter, progressing towards an "Omega Point" that is the culmination of the whole evolutionary process. At that point, matter becomes wholly infused with spiritual energy and all phenomena, including human beings, attain oneness. Teilhard de Chardin believed that the increasing interconnection of the human race was a part of this evolutionary process, and eventually our species would unite into a single interthinking group. (Teilhard de Chardin died in 1955, so it's interesting to ponder over what he would have made of the age of the Internet, and all of the increasing interconnection it has brought about. No doubt he would see it as part of our species' movement towards oneness.)

There is also a parallel between the panspiritist view of evolution and the concept of emergence. Systems theorists suggest that natural systems and organisms have an innate tendency to move towards greater order and complexity, spontaneously generating structures which are more than the sum of their parts. As we saw earlier, biologists such as Robert Reid and Stuart Kauffman have applied the concept of emergence to evolution as an alternative to the Neo-Darwinist model. According to this view, order and complexity are not created by genetic mutations but by self-organization and the emergent properties of systems. Robert Reid was doubtful about *any* role for natural selection in evolution, whereas Stuart Kauffman believes that this spontaneous tendency to order works alongside natural selection.

The only real difference between this theory and spiritual evolution is that the former suggests that emergence happens spontaneously, as some kind of inherent property of systems, and of life itself. Of course, this doesn't explain where this inherent property comes from, or even what it is. But the spiritual evolution theory provides a little more explanation: this tendency towards complexity stems from consciousness itself, which generates greater complexity in order to support its intensification.

Adaptative mutation – non-random mutations?

This innate tendency towards complexity helps to explain a strange aspect about evolution – its *creativity*.

The palaeontologist Simon Conway Morris has written of "The uncanny ability of evolution to navigate to the appropriate solution".[16] One example of this is the way that life forms sometimes adapt to changes in their environment with a rapidity that would be impossible through random genetic mutations. The technical term for this is "adaptive mutation". In these cases, mutations sometimes occur in a specific response to environmental challenges or stresses, such as changes in temperature, in nutrients or population size. For example, research has found that if a strain of bacteria is unable to process lactose and it is then placed in a lactose-rich medium, 20 per cent of its cells will quickly mutate into a Lac+ form and become able to process the lactose. The mutations become part of the bacteria's genetic code, and are inherited by following generations.[17] In adaptive mutation, it's almost as if the mutations aren't random at all but are somehow being "directed" to react to the situation in the appropriate way, exactly when required.

One suggested explanation for adaptive mutation comes from the relatively new field of "quantum biology", which we briefly looked at in the last chapter (and will look at in more detail in Chapter 11). Quantum biology attempts to explain mysterious biological phenomena in terms of the principles of quantum physics, such as superposition and entanglement. Applied to the example just mentioned, a quantum explanation would be that the genome of the bacteria exists in a state of "superposition". That is, it doesn't exist in any one particular state, but in a myriad of possible states, some of them mutated and others non-mutated. But when certain circumstances arise, the genome "collapses" into the appropriate mutated state.[18]

However, adaptive mutation could simply be an expression of the same creativity that allows life forms to move towards greater complexity and consciousness. This creativity gives life forms the flexibility to respond to challenges. There is a dynamic quality that enables them to develop in the appropriate way.

This also suggests that a spiritual view of evolution doesn't have to dispense with genetic mutations as an important factor. Genetic mutations may still be the main overt way in which change occurs. The only difference is that, according to this view, beneficial mutations don't happen (or at least don't always happen) randomly. They may be generated by the impetus of evolution, as a means of creating change. From this perspective, mutations occur as a part of the unfolding of the process of evolution, generating inevitable changes that lead to more complex and conscious forms.

Co-operation and competition

Another flawed aspect of the standard view of evolution is its emphasis on competition. Evolution is usually conceived as a kind of war, a "survival of the fittest" in which "genetic machines" ruthlessly fight against one another for access to resources, and relate to one another as rivals and threats. This probably stems from the great emphasis that Neo-Darwinists put on natural selection as the main "filtering" mechanism in evolution. In a hypothetical sense, natural selection would work best in an environment of extreme competition. So Neo-Darwinists may have unconsciously chosen to conceive of evolution in these terms in order to make natural selection seem more feasible as an explanation for evolutionary change.

Darwin himself didn't see evolution in quite the same way. (As noted earlier, it was TH Huxley who coined the phrase "survival of the fittest".) Darwin was fully aware of the co-operative aspects of evolution, and gave many examples of them in *On the Origin of Species*. (This was probably because he didn't believe that natural selection was the only factor in evolution, and so didn't need to emphasize it so much.) Darwin did use the term "struggle for existence", but stated that he used the term "in a large and metaphorical sense, including dependence of one being on another".[19] In other words, it was a collective struggle, not an individual one, which relied on what he called a "web of complex relations" between different organisms, with a great deal of interaction and co-operation, rather than competition alone.

However, in the materialistic fervour of the last decades of the 19th century, scientists and intellectuals like TH Huxley and Herbert Spencer emphasized the competitive and individualistic aspects of Darwin's theory. The theory was distorted into the popular image of evolution as a violent and chronic struggle for limited resources. This distortion of the theory was even applied to human affairs, in the form of "Social Darwinism", which viewed human society as an individualistic struggle for wealth and status, and helped to form the philosophy of modern Western capitalism. And ever since, scientists like Richard Dawkins have continued to exaggerate the competitive aspects of Darwin's theory and to ignore its co-operative aspects.

One early dissenter to the theories of Social Darwinism was the Russian scientist Peter Kropotkin, who was living in exile in London at the end of the 19th century, after being imprisoned in Russia for his political activism. As an anarchist-communist, Kropotkin was horrified at how English intellectuals were applying Darwin's theory to justify the excesses of capitalism. He admitted "there is [an] immense amount of warfare and extermination going on amidst various species", but he added: "there is, at the same time, as much, or perhaps even more, of mutual support, mutual aid, and mutual defense... Sociability is as much a law of nature as mutual struggle."[20] In his 1902 book *Mutual Aid*, Kropotkin gave many examples of co-operation among animals, mostly focusing on the co-operative swarm behaviour of insects and other species, such as when beetles work co-operatively to bury the corpses of larger animals such as mice or birds, and how land crabs travel in large groups to the sea to deposit their spawn.

Such co-operation and altruism between members of the same groups could easily be explained in terms of the concept of "kin selection". This simply means that by helping other members of their own groups, insects or animals are indirectly supporting the survival of their own genes because the other members are genetically so closely related to them. However, co-operation and altruism can also cross the boundaries between species. In biology this is called mutualism, or symbiosis. Examples of this include tubeworms which have no guts but manage to survive with the help of bacteria that digest food for them. In return, the tubeworms provide them with hydrogen sulphide and methane. Another example is clownfish, which live in a symbiotic relationship with sea anemones, helping each other in a variety of ways: for example, fecal mater from the clownfish gives nutrients to the sea anemone, while the anemone's stinging cells help protect the clownfish from predators. There are also many lichens, corals and plants that "allow" bacteria or fungi to live inside them, and in return receive essential nutrients.

Co-operation even occurs at the level of genes. One of the notions that the Human Genome Project (which I mentioned briefly in Chapter 2 and will discuss again in the last chapter of this book) helped to dispel was the image of genes as discrete individual entities with specific roles, which fight with each other for survival, like gladiators in a ring. The project showed that the reality was much more complex than this, and that most genes co-operate and multitask. Single genes can code for several different proteins, and may have other roles too. (As noted earlier, another myth that was dispelled by this was the popular belief that there are "genes for" certain traits and characteristics.) In reality, the survival of all genes depends on co-operation. The genome isn't like a swimming pool full of individual competitors in a race against each other. It's more like a pool where a water polo or synchronized swimming team are practising, working together and co-ordinating all their activities for the common good.

Some biologists – such as Lynn Margulis and James Lovelock – have argued that co-operation is more important than competition in evolution, and is in fact its real driving force. Living beings do not survive by fighting against one another, but by interaction and mutual dependence. If the phrase "the survival of the fittest" has any meaning, it means the survival of those who interact most effectively. As Margulis (with her co-author Dorion Sagan) has written, "Life did not take over the globe by combat, but by networking. Life forms multiplied and complexified by co-opting others, not just by killing them."[21] This applies to one of the most important steps in the whole of the evolution of life on this planet, when prokaryote cells (the simplest single-celled entities, with no nucleus) came together to form eukaryotes (unicellular or multi-cellular organisms that do have a nucleus). This didn't happen through competition, but through co-operation and co-ordination.

This perspective makes little sense in terms of the standard Neo-Darwinist model, which is why the significance of co-operation is often downplayed. But in terms of the panspiritist model, co-operation makes complete sense. One of the fundamental assumptions of Neo-Darwinism – and of materialism in general – is that living beings are separate and distinct entities, whose only concern is for the survival and replication of their genes. But this is a pernicious falsehood. In reality, there is no separation, and not even any individuality. All living beings are interconnected. We are all expressions of the same spiritual force, no matter how different we are physically. In the same way that water can take many different forms – rain, lakes, ponds, streams, rivers, oceans – but is always essentially the same, consciousness expresses itself in many different forms, but is always essentially the same. The same spiritual energy appears as the essential being of all living creatures. It's therefore not at all surprising that living beings don't exist in a state of wholesale warfare with each other, but instead exhibit a great deal of co-operation and altruism.

Evolutionary psychology

One of the most regrettable applications of the idea of evolution as a individualistic struggle for survival – or a competition for limited resources – is the field of "evolutionary psychology". This field attempts to apply Neo-Darwinism to human behaviour. Its basic idea is very simple: if a human characteristic or trait has survived to the present day, then it must have had some evolutionary advantage for our ancestors. Traits are obviously linked to certain genes, and those genes could only have been "selected" if they helped our ancestors to survive in some way, or if they conferred a reproductive advantage.

Take the question of why many people – particularly men, it seems – apparently have a strong desire to attain wealth, power and status. According to evolutionary psychologists, this is because, in prehistoric times, wealth and power would have enhanced people's chances of survival. At the same time – particularly for men – wealth and power would have increased their reproductive possibilities. Women would have been attracted to wealthy and powerful men, because they offered greater security. (This, incidentally, also explains why societies tend to be male-dominated, because men have a stronger desire than women for power and wealth.)

The apparent selfishness of human beings can also be explained in these terms. Since life was extremely hard in prehistoric times, the people who were most ruthless and least compassionate were more likely to hold on to food and resources themselves, and therefore more likely to survive. Compassion and sharing would have probably decreased the individual's chances of survival. Even more questionably, some evolutionary psychologists have attempted to explain rape as an evolutionary adaptation. According to this logic, because rape is common in most societies there must be some evolutionary reason for its existence. So the theory has been put forward that rape is a desperate attempt by low-status men, who cannot attract willing sexual partners, to replicate their genes.[22] (In the next chapter, we will also look at evolutionary psychology's explanations of other unsavoury traits such as racism and war.)

There are so many problems with these explanations – and with the field of evolutionary psychology as a whole – that it's impossible to address them all here.[23] But one of the questionable aspects of the theory is the assumption that all present-day human traits must have had some adaptive value for our ancestors (or else be a by-product of other adaptive traits). There are many prevalent human traits where it is difficult to perceive any survival value. For example, take depression; according to the logic of evolutionary psychology, because depression is so common, it must have had some adaptive value. Hence some evolutionary psychologists have suggested that depression was an adaptation that evolved when our ancestors were faced with complex problems and developed the ability to ruminate over them.[24] However, this is really just a post hoc argument based on a pre-existing assumption. It only makes any sense if one assumes that there must be an evolutionary reason for depression. If one doesn't assume this, then the explanation seems quite bizarre, for

a number of reasons: for example, the lack of evidence for
depression in hunter-gatherer groups; the probable lack of
necessity for extended rumination of this kind in prehistoric
life; and the sheer life-negating effects of depression, which
must surely outweigh any possible benefits. Severe depression
may lead to suicide, the least desirable of any evolutionary
outcome. Surely any trait that encourages the impulse for self-
destruction would not have been selected by evolution.

Some evolutionary psychologists have also used this kind of
argument to explain the evolution of consciousness. Because
we are apparently conscious, consciousness presumably must
have had some adaptive value. One theory is that consciousness
has had survival value because it convinces us that we are
special, unique individuals and so makes us more determined
to survive. But why should the feeling of being a "special self"
be advantageous when many other living beings survive well
enough without (apparently) possessing it? Another theory
is that consciousness gave us the ability to understand other
people, to work out what they might be thinking or planning.
We became able to "second guess" or "outguess" them and this
helped us in the competition for survival.[25]

However, consciousness could just as easily be seen as a
disadvantage in survival terms. For example, it could be argued
that it makes us liable to suffer from anxiety, frustration and
self-hatred. Our consciousness of ourselves and of our lives
(and our potential death) means that more human beings
commit suicide than members of any other species. In
addition, as the German philosopher Thomas Metzinger has
pointed out, on a wider scale, consciousness could be seen as
maladaptive, because we as human beings can't seem to live in
harmony with the Earth, and are in danger of destroying the
life-support of our planet, and hence killing ourselves.[26]

The effort to explain human traits in terms of their
evolutionary benefit is really just a kind of dogma that leads
to absurd and offensive justifications of behaviours, such as
rape and male domination. It's not too dissimilar from the
way that religious people try to explain every aspect of human

life with reference to God, including problematic aspects like the existence of pain and suffering. In reality, evolutionary adaptation is just one out of a whole host of factors (and probably not one of the most important ones) which influence human behaviour, including social or environmental, psychological, existential, spiritual and accidental factors.

Evolutionary psychologists tend to pick certain aspects of what they believe is "human nature" and create "just so" stories to explain their development, based on the supposed benefits these traits would have had in early human history. One fallacy of this is that "human nature" is extremely nebulous and can be interpreted in a variety of ways. Who says that human beings are innately selfish? That men have a strong desire for wealth and power? And that women are generally attracted to wealthy and powerful men? This is just a "cherry-picked" view of human nature, made up of the elements which adhere to the Neo-Darwinist view of life as a struggle and a competition. As we'll see in the next chapter, just like other species, human beings are often collaborative rather than selfish and competitive. We are just as likely to be benevolent as we are to be ruthless. There are many racist people, but there are also many people who feel empathy and inclusiveness towards other ethnic groups. Some women might be attracted to wealthy and powerful men, but many others are attracted to caring, sensitive men, irrespective of their social status.

However, perhaps the most serious problem with evolutionary psychology is that it is based on a completely false image of the human race's past. It is based on the Neo-Darwinist assumption that – like evolution as a whole – human history (and prehistory) has been a brutal struggle for survival. The early period during which human traits developed is viewed as a time when human life was "nasty, brutish and short", and only the traits which gave people a survival advantage were selected, while all others fell by the wayside. In the next chapter, we will see that this is a crude caricature.

In view of these issues, it seems strange that evolutionary psychology has become such a popular theory. But this is probably because – like materialism in general – it has a great deal of explanatory power. It offers a very simple way of accounting for all aspects of human behaviour. Another reason for its popularity is that it fits so well with the competitive and individualistic values of modern Western societies. Like the general notion of life as a "survival of the fittest", evolutionary psychology was born out of these values. The picture of early human life as a struggle for genetic success, with individuals and groups competing for access to limited resources, is a good metaphor for competitive capitalist societies. And since this is the kind of society we are familiar with, the theory seems to make sense to us.

However, a more egalitarian culture might well have come up with a more collaborative and benevolent model of human behaviour – and as we will see in the next chapter, evidence would certainly have been found to justify this view.

An alternative view of evolution

Just to summarize then, the spiritual view of evolution suggests that there is an impulse in consciousness itself to express itself more intensely within life forms, and to generate more complex forms of life in order to support greater intensities of awareness.

I sometimes compare evolution to the development of a human being from embryo to adulthood. Development moves naturally and inevitably from the simplest state (when two cells meet and merge) through levels of increasingly complexity, as cells split off and organize and start to form different parts of the body. The process unfolds along predetermined lines, following a kind of blueprint or mould that is specific to our species. I think the process of evolution is similar to this, but on a massively extended time frame, unfolding over hundreds of millions of years. Perhaps the only difference is that the direction of evolution may not be as fixed as the development of individuals – perhaps there is a simple tendency to move towards greater complexity and awareness, which is broadly directional without being completely predetermined.

So, in my view, to believe that the process of evolution is accidental is as illogical as interpreting human development from embryo to adulthood as a random process. The process of ontogenetic (or individual) development closely parallels the course of evolution itself over the past four billion years, moving from simple cellular structures to increasing complexity and specialization – and this parallel includes the probability that both types of development are not random, but directional.

One of the interesting aspects of this view of evolution is how it connects with spiritual development. In my book *The Leap*, I suggested that when people have "awakening experiences", or when they experience an ongoing state of wakefulness, they are glimpsing a higher stage of evolution. These are experiences of a more intense and more expansive awareness, and in this sense they represent a continuation of the expansion of awareness that has taken place since the beginnings of life roughly four billion years ago. Spiritual development is a process of self-evolution and it relates to the evolutionary process as a whole. The impulse that many people feel to develop themselves spiritually – that is, to expand and intensify their awareness – is essentially the same impulse that has taken life forms from the simplest single-celled prokaryotes to multibillion-brain-celled creatures like dolphins, whales and human beings. Consciousness is working through us, moving us towards a more intense form of awareness (and perhaps, as Pierre Teilhard de Chardin believed, this is related to the increasing intra-species interconnectivity created by modern trade, transport and communication – and especially the Internet). One day, what we presently know as a "higher" state of consciousness will become normal (to other living beings, if not human beings). Then what we presently experience as "normal" awareness will seem as limited as the awareness exhibited by simpler forms of life such as insects appears to us now.

CHAPTER 10

WHY DO SELFISH GENES BEHAVE SO UNSELFISHLY? THE PUZZLE OF ALTRUISM

On 22 May 2017 my home city of Manchester, England, suffered a terrorist attack. A 22-year-old man waiting in the foyer at the end of a concert by Ariana Grande detonated a bomb strapped to his chest, killing 22 people (including himself) and injuring over 500. Most of the victims were either children or parents waiting to collect their children. However, in the midst of the senseless savagery of the attack there were many stories of heroism and selflessness.

An off-duty doctor who was walking away from the concert after picking up his daughter ran back into the foyer to help the victims. A woman who saw crowds of confused and frightened teenagers running out of the venue guided around 50 of them to the safety of a nearby hotel. There she shared her phone number on social media so that parents could come and pick their children up. Taxi drivers across the city switched off their meters and took concertgoers and other members of the public home. Taxi drivers from as far as 30 miles (48 kilometres) away converged on the city to offer free transport. A homeless person named Stephen Jones was sleeping rough near the venue and rushed in to help. He found many children covered with blood, screaming and crying. He and a friend pulled nails out of the children's arms – and, in one case, out of a child's face – and helped a woman who was bleeding severely by holding her legs in the air. "It was just my instinct to go and help people out," he said.[1] (Although – to illustrate the negative side of human nature – another homeless man was convicted of stealing belongings from the injured victims of the attack.) As one paramedic – named Dan Smith – who was at the scene commented, "There was an unbelievable amount of people doing what they could to help… I saw people pulling together in a way I have never seen before… The thing I will remember more than any other is the humanity that was on display. People were catching each other's eye, asking if they were okay, touching shoulders, looking out for one another."[2]

Such acts of altruism are almost always a feature of emergency situations. Also in the UK, in 2016 a cyclist was trapped under the wheel of a double-decker bus. A crowd of around 100 people gathered together, and in an amazing act of co-ordinated altruism, lifted the bus so that the man could be freed. According to a paramedic who treated the man, this was a "miracle" that saved his life. Another example took place in Glasgow when a helicopter crashed into a pub in November 2013, killing ten people. Soon after the crash, residents and passers-by rushed towards the scene. Together with some of the pub's clientele, they formed a human chain, passing wounded and unconscious victims, inch by inch, out of the danger area and into the hands of the emergency services.

As one final example, in 2007 a construction worker named Wesley Autrey was standing on a subway platform in New York when a young man nearby had an epileptic seizure and rolled onto the track. Hearing the approach of a train, Autrey impulsively jumped down to try to save the young man, only to realize that the train was approaching too fast. Instead, he jumped on top of the young man's body and pushed him down into a drainage ditch between the tracks. The train operator saw them, but it was too late to stop: five cars of the train passed over their bodies. Miraculously, both of them were uninjured. Asked later by *The New York Times* why he had done it, Autrey said: "I just saw someone who needed help. I did what I felt was right."[3]

The cold "truth"

These examples demonstrate that although we human beings can sometimes be selfish and competitive, we can also be extraordinarily kind and selfless. However, the materialist worldview tends to downplay the benevolent aspects of our nature, and even explain them away. Capitalist economic systems – derived from the materialist worldview – encourage us to compete with others to gain success and wealth, and to see our fellow human beings as rivals. And as we saw in the last chapter, the theories of Neo-Darwinism and evolutionary psychology portray human beings as ruthless genetic machines, only concerned with survival and reproduction.

One of the most influential books of the second half of the 20th century was Richard Dawkins's *The Selfish Gene*, which – as with the field of evolutionary psychology in general – became popular because it seemed to offer scientific confirmation and justification of the ruthless individualism of Western societies. And in a passage from the book, Dawkins expresses the "cold truth" about life according to Neo-Darwinism:

> To a survival machine, another survival machine (which is not its own child or another relative) is part of its environment, like a rock or a river or a lump of food. It is something that gets in the way, or something that can be exploited. It differs from a rock or a river in one important respect: it is inclined to hit back. This is because it too is a machine that holds its immortal genes in trust for the future, and it too will stop at nothing to preserve them. Natural selection favours genes that control their survival machines in such a way that they make best use of their environment. This includes making the best use of other survival machines, both of the same and of different species.[4]

This passage is shocking in its brutality. It portrays human beings as psychopathic predators in a similar way to extreme right-wing ideologies such as Nazism. Dawkins would probably say that he is simply "telling is like it is", and in a sense this is true; he's simply taking the materialist perspective to its logical conclusion. If we are nothing more than "carriers" of thousands of genes, whose only aim is to survive and replicate themselves, then of course we (like all other living beings) are selfish and ruthless. (In fairness to Dawkins, he is not an apologist for psychopathy or fascism – he believes that we should accept the fact that we are fundamentally selfish and brutal, but try to control and curtail these impulses.)

The problem is that, as the previous examples show, there are frequent occasions when we human beings don't behave at all like ruthless predators – when, in fact, we behave in precisely the opposite way and sacrifice our own well-being (potentially even our own lives) for the sake of others. If we are only interested in our own survival, this behaviour doesn't make sense.

This is the basic issue we're going to discuss in this chapter: if living beings are made up of selfish genes, why do they often behave so unselfishly? Of course, materialism and Neo-Darwinism do have some explanations to try to account for altruism, and we will examine these. We will see that these are highly problematic, and largely based on erroneous assumptions and questionable data.

Prehistoric altruism

As we saw in the last chapter, the idea that human beings are innately selfish is based on the idea that life is fundamentally a struggle for survival. According to the standard narrative, selfishness and ruthlessness have flourished because we have always had to compete for limited resources. Altruism would not have benefitted us at all; it would have meant giving up access to resources that we need for our own survival. But selfishness would have increased our chances of survival, and so it would have been "selected" by evolution as a trait, and as a result passed down to future generations.

This applies to both individuals and groups. As individuals within groups, men ruthlessly sought wealth and power so that they could ensure their own survival and become more attractive to the opposite sex, and as a result propagate their genes as widely as possible. And at a group level, tribes (usually made up of genetically closely related individuals) competed against each other for resources. Other tribes were a potential threat to our survival, because they might use up the resources we needed, and so it was inevitable we were hostile towards them.

Some evolutionary psychologists see this as the origin of racism and warfare. One group would inevitably be hostile towards others, depriving them of access to resources, in order to increase their own access to them and enhance their chances of survival. In the words of anthropologist Pascal Boyer, racism is "a consequence of highly efficient economic strategies", enabling us to "keep members of other groups in a lower-status position, with distinctly worse benefits".[5] And in terms of warfare, conflict inevitably arose between different groups competing for access to the same resources, such as food and water.

But this is a crude caricature of prehistoric human life. Back then, life was not a struggle. Archaeological and anthropological evidence has shown that our ancestors had a good diet, a lot of leisure time and were free from most of the diseases that afflict modern humans (many of which were passed on to us later by the animals – such as cows, sheep and horses – we domesticated following the switch to a farming lifestyle).[6] And life was also easy in the sense that our ancestors didn't have to compete against other groups for resources. In the prehistoric era, the world was very sparsely populated, which meant there was actually an abundance of resources for hunter-gatherer groups. According to some estimates, around 15,000 years ago the population of the whole of Europe was less than 30,000, and the population of the whole world was no more than half a million.[7] Population densities were therefore very low, with an absence of the kind of survival pressure that would have favoured traits like competition and selfishness. In evolutionary terms there was no reason why these traits should have been selected as adaptations.

It is also highly unlikely that groups ever became so large that they would exhaust the resources of a particular area, because they moved sites regularly – usually every few months – and experienced very little population growth. Before the advent of agriculture, rates of population growth were extremely slow – well below 0.001 per cent per year, according to one estimate.[8] Possible reasons for this are longer periods of breast-feeding (up to the age of five or six, resulting in extended periods of infertility) and the use of plant contraceptives. (Certainly, a small number of children is more suitable for the mobile hunter-gatherer lifestyle because larger numbers become more difficult to transport to new sites.)

The inaccuracy of evolutionary psychology's view of prehistory is also shown by a lack of evidence for warfare in prehistoric times. I presented a lot of data about this in my book *The Fall* – published in 2005 – but since then a great deal more evidence for a state of "prehistoric peace" has accumulated. For example, in 2013 the anthropologists Jonathan Haas and Matthew Piscitelli surveyed descriptions of 2,900 prehistoric human skeletons from scientific literature. Apart from a single massacre site in Sudan (in which two dozen people were killed), they found only four skeletons that showed signs of violence – and even these signs were consistent with homicide rather than warfare.[9] This dearth of violence completely contrasts with later periods when signs of war become obvious from skeletal marks, weapons, artwork, defensive sites and architecture. In the same year, another anthropologist, Brian Ferguson, carried out a detailed survey of archeological findings from Neolithic Europe and the Near East, which found almost no evidence of warfare.[10]

Even contemporary hunter-gatherers – at least those who still live the same lifestyle as our prehistoric ancestors – are generally unwarlike.[11] In 2014 a study of 21 contemporary hunter-gatherer groups by the anthropologists Douglas Fry and Patrik

Söderberg found a striking lack of evidence for inter-group conflict over the last 100 years. There was only one society (an Australian Aboriginal group called the Tiwi) who had a history of group killings.[12]

It's likely that in prehistoric times there was very little competition *within* individual groups either. Anthropological reports of hunger-gatherer groups who live the same "immediate return" way of life as our ancestors — meaning that they consume their food almost straight away, without storing surpluses — have always shown them to be extremely egalitarian and democratic. Most groups do not have leaders or hierarchies, and they reach decisions by consensus. They do not hoard goods or collect possessions, and have very strongly developed practices of sharing. This equality of status applies to women too. Recent research on contemporary hunter-gatherer groups has shown that men and women tend to have equal status and influence, leading to the suggestion that sexual inequality was a relatively recent social development.[13]

In other words, the idea of life as a competition for limited resources, and as a general struggle for survival, has no meaning in terms of the mobile hunter-gatherer lifestyle our species has followed for the great majority of our time on this planet. And so the idea that human beings are naturally selfish and competitive is predicated on a false narrative, and is therefore itself a fallacy.

In reality, life only became a struggle in relatively recent times, once our ancestors had settled down and taken up farming. That was when resources became more scarce (partly due to increased population densities). And that was also when we were obliged to work much harder to attain food, when we became more susceptible to illness and when our diet deteriorated. This is why many anthropologists and archaeologists regard agriculture as — in the words of Jared Diamond — "the worst mistake in the history of the human race".[14]

If anything, judging from the way that our ancestors actually did live, altruism and co-operation should be much more innate to us than competition and aggression. And I believe that this is actually the case. The fact that, in crisis situations, people respond so instinctively in an altruistic way, without any conscious deliberation, illustrates this.

Explanations for altruism

It's true that, in genetic terms, it's not necessarily self-defeating for us to help people close to us – our relatives or distant cousin carry many of the same genes as us, and so helping them may help our genes to survive. As Richard Dawkins explains, "altruism at the level of the individual organism can be a means by which the underlying genes maximize their self interest."[15] So what may appear to be self-sacrifice may actually mean perpetuating our own genes.

But what about when we help people who have no relation to us? Or when we help animals?

Materialists have suggested a variety of different explanations to account for this. According to some psychologists, there is no such thing as "pure" altruism. When we help strangers (or animals) there must always be some benefit to us, even if we're not conscious of it. Altruism makes us feel good about ourselves; it makes other people respect us more; or it might (for religious people) increase our chances of getting into heaven. Or perhaps altruism is an investment strategy – we do good deeds to others in the hope that they will return the favour some day, when we are in need. (This is known as reciprocal altruism.) According to evolutionary psychologists, altruism could even be a way of demonstrating our resources, showing how wealthy or able we are so that we become more attractive to the opposite sex, which enhances our reproductive possibilities.

Finally, evolutionary psychologists have suggested that altruism towards strangers may be a kind of mistake, a "leftover" trait from when human beings lived in small groups with people we were genetically closely related to. Of course, we felt an instinct to help other members of our group because our own survival depended on the safety of the group as a whole, and because, more indirectly, this would support the survival of our genes. So even though we don't live in small tribes of extended families anymore, we instinctively behave as if we do, helping the people around us as if we are related to them.

What all these explanations have in common is that they are really attempts to explain *away* altruism, or to make excuses for it. It's almost as if we're saying, "Please excuse my kindness, but I was really just trying to look good in the eyes of other people." Or perhaps, "Sorry for helping you, but it's a trait I picked up from my ancestors thousands of years ago, and I just can't seem to get rid of it." Obviously these reasons apply sometimes. Many acts of kindness may be primarily – or just partly – motivated by self-interest. But "pure" altruism surely exists too – a simple, direct desire to alleviate the suffering of other human beings, or other living beings, based on our ability to empathize with them. An act of pure altruism may make a person feel better about themselves afterwards, and it may increase other people's respect for that person, or increase their chances of being helped in return at a later point; but it's possible that at the very moment when the act takes place their only motivation is an impulsive, unselfish desire to alleviate suffering.

Empathy as the root of altruism

The other day, I was about to have a shower when I saw a spider near the plughole of our bath. I got out of the shower, found a piece of paper, gently encouraged the spider onto it, and scooped it out of danger.

Why did I do this? Perhaps in the hope that a spider would do the same for me in the future? Or that the spider would tell his friends what a wonderful human being I am? Or, more seriously, perhaps it was the result of moral conditioning, a respect for living things and an impulse to "do good" that was ingrained in me by my parents? (Although come to think of it, my parents didn't actually teach me those things...)

I'm being a little facetious, but the question of altruism towards members of other species is an important one, since it can't be explained in genetic terms, or in terms of "reciprocal altruism". If I donate money to an animal charity, stop to pick up an injured bird on the road and go ten miles (16 kilometres) out of my way to take it to the nearest vet, am I really doing it to look good in other people's eyes, or to feel good about myself? Again, that *could* be the case, but it's also possible that these are acts of pure altruism – responses to the suffering of another living being, arising out of empathy. It is possible that I simply empathized with the spider as another living being, one that was entitled to stay alive just as I was.

From the panspiritist perspective, empathy is the root of all pure altruism. Empathy is sometimes described as the ability to see things from another person's perspective, or "put yourself in their shoes". But in its deepest sense, empathy is the ability to *feel* – not just to imagine – what others are experiencing. It's the ability to actually enter the "mind space" of another person (or being) so that you can sense their feelings and emotions.

In this way, empathy is the source of compassion and altruism. Empathy creates a connection that enables us to feel compassion. We are able to sense the suffering of others and this gives rise to an impulse to alleviate their suffering, which in turn gives rise to altruistic acts. Because we can "feel with" other people, we are motivated to help them when they are in need.[16]

Connection through altruism

Altruistic acts are beneficial not just to those who receive them, but also to those who perform them. Generosity is strongly associated with well-being. Studies of people who practise volunteering have shown that they have better psychological and mental health and increased longevity. The benefits of volunteering have been found to be greater than taking up exercise, or attending religious services – in fact, even greater than giving up smoking. One study found that when people were given a sum of money they gained more well-being if they spent it on other people, or gave it away, rather than spending it on themselves.[17]

Every year, on the Positive Psychology module I teach at my university we have a session on empathy and altruism. After the session, I encourage the students to go away and perform as many acts of kindness as they can over the following week, and then reflect on how the acts made them feel. Typical acts of kindness they perform are stopping to talk to homeless people or buying them food, offering to do chores for elderly neighbours (or simply spending time with them), giving lifts to friends or even simple acts like holding doors open for people. Almost without fail they report back that these altruistic acts gave them a strong sense of well-being, made them feel positive about themselves, about other people and life in general.

Why does altruism give us such a strong sense of well-being? I believe this is more than simply "feeling good" about ourselves, or showing off to others. This sense of well-being comes from the powerful sense of empathic connection that is associated with altruism. A sense of connection generates the impulse to be altruistic, and when we perform an act of kindness, the sense of connection intensifies further, bringing about a transcendence of self-centredness and separateness.

Sometimes this sense of connection is experienced very directly during an altruistic act. This was commented on by some of my students. One of them described a strong sense of connection she felt to a homeless person, writing: "I felt a real bond with him, as if I could see what life was really like for him and felt respect for him. A lot of people objectify the homeless, and think of them as having substance abuse issues or psychiatric problems, but it made me realize that they are just as human and as sensitive and as deserving of good things as anyone else." Another student described a spiritual sense of connection that went beyond the elderly person she was helping: "It was not just a simple sense of connection with her, although it definitely was that as well. It was like plugging into something bigger that everyone is a part of."

Elation

Interestingly, the positive effects of altruistic acts – including the sense of empathic connection and well-being they bring –

can extend beyond the performer of the act and its recipient. An acquaintance of mine used to live in Pakistan, where he was an animal rights activist. One day, he was walking through his home city when he saw a crowd gathered around the stall of a birdseller. A man had bought some myna birds – a popular caged bird in Pakistan, because of their ability to mimic sounds – and was releasing them. One by one, he took them out of the cage and let them fly free. In all, he bought 32 of the birds, just to set them free.

My friend was amazed by this act of altruism, partly because – as he put it – "such acts of charity were not so common in my part of the world where people are not so kind to animals in general". But he was also amazed at his own reaction to the act. He was filled with a deep sense of peace. A strange quietness filled his surroundings and he felt completely free of worry or anxiety. The sense of peace and joy remained with him for a few days, and, in his words, "I believe it is still there to some degree."

I would guess that most of you have had a similar – if perhaps not as intense – experience. It's the fantastic warm, elevated feeling we get when we witness acts of kindness. Even the simplest altruistic acts might give you a touch of this feeling: a passer-by giving his packed lunch to a homeless man; a stranger offering to help a visually impaired person cross the road; or a subway passenger giving up his seat for an elderly person.

In this way, witnessing altruistic acts can be a source of what Abraham Maslow called "peak experiences" – those moments of awe, wonder and a sense of "rightness" that make us feel immensely grateful to be alive. The psychologist Jonathan Haidt calls this experience "elation" and describes it as a "warm feeling in the chest, a sensation of expansion in [the] heart, an increased desire to help, and increased sense of connection with others". It is also, in his words, "a manifestation of humanity's 'higher' or 'better' nature".[18]

In fact, this may be one reason why the experience of elation occurs – because it brings a renewed faith in human nature, a sense of the sheer goodness that human beings are capable of, although this might sometimes be difficult to see amidst the chaos and conflict of everyday life. But, as Haidt himself suggests, a sense of empathic connection is a very important aspect of elation too. In these moments we become part of a shared network of being, and feel our essential oneness. Altruistic acts often have a three-way effect: they generate a sense of connection (and therefore an intense sense of well-being) for the recipient, the performer and any witnesses.

The source of altruism

In panspiritist terms, therefore, altruism is easy to account for: it's simply the result of empathy. Our capacity for empathy shows that, in essence, all human beings – and in fact all living beings – are interconnected. We are expressions of the same consciousness. We share the same essence. We are waves of the same ocean, influxes of the same all-pervading spiritual energy.

It's this fundamental oneness that makes it possible for us to identify with other people, to sense their suffering and respond to it with altruistic acts. We can sense their suffering because, in a sense, we *are* them. And because of this common identity, we feel the urge to alleviate other people's suffering – and to protect and promote their well-being – just as we would our own. And, as in the examples in the previous section, it's this fundamental oneness that we actually *experience* – as a feeling of connection – when we perform (or witness or receive) altruistic acts.

This relationship between altruism and our fundamental oneness was expressed beautifully by the 19th-century German philosopher Arthur Schopenhauer, who wrote: "My own true inner being actually exists in every living creature, as truly and immediately known as my own consciousness in myself... This is the ground of compassion upon which all true, that is to say unselfish, virtue rests, and whose expression is in every good deed."[19] Or in the words of the Spanish Jewish mystic Moses ben Jacob Cordovero: "In everyone there is something of his fellow-man. Therefore whoever sins injures not only himself but also that part of himself which belongs to another."[20] In this way, according to Cordovero, it is important to love others because "the other is really oneself".

In other words, there is no need to make excuses for altruism. Instead, we should celebrate it as a transcendence of seeming separateness. Rather than being unnatural, altruism is an expression of our most fundamental nature – that of oneness.

CHAPTER 11
QUANTUM QUESTIONS: MYSTERIES OF THE MICROCOSM

The macrocosmic physical world we see as we go about our lives seems to make sense. It seems to obey fundamental natural laws. Objects move through space and time according to the laws of motion described by Newton. From this standpoint, the world appears to be a kind of machine, the functioning of which can be understood. And scientists used to assume that the microcosmic world of atoms followed the same laws as the macrocosmic world, and could be understood in the same way. This seems to be a logical assumption – after all, the closer you look into a machine the more mechanical detail you find. And to a certain point, this seems to hold true. For example, the behaviour of molecules seems to be fairly mechanical and comprehensible.

But beyond this, everything seems to go haywire. When we look deeper than the level of the molecule, things don't become more mechanical, but more counterintuitive and mysterious.

This is the world of quantum physics, which investigates the behaviour of atomic and sub-atomic particles. The strange thing about this world is that it is so different to the macrocosmic world. The laws that seem to work so well in the everyday world are flatly contradicted in the quantum world. The picture of the world we gain from classical Newtonian physics – and which forms the basis of the materialist worldview – has so little in common with what quantum physics tells us is the reality of things that they could be describing two different universes rather than the same one.

In this sense, quantum physics has always had an uneasy relationship to the rest of modern science. In a way, it has always been an "enemy within", at least as far as materialism is concerned. Decades before "anomalous" areas of modern science such as neuroplasticity and the placebo effect, quantum physics undermined the basic principles of materialism.

We've already touched on quantum physics several times throughout this book, most notably in Chapter 8, in relation to psi phenomena. There we looked at the concept of time, and found that in the sub-atomic world time isn't necessarily linear, which allows for the possibility of precognition. We also saw that the concept of "entanglement" allows for telepathy. In this chapter we're going to look at the area in more detail, and to examine more deeply its problematic relationship to materialism.

However, a word of warning before we begin: you may find some of the material we're going to look at difficult to understand; it may seem more like science fiction than reality as we know it. But don't worry – as the 20th-century physicist Richard Feynman is believed to have said, "If you think you understand quantum mechanics, you don't understand quantum mechanics." And don't worry if you find some of the material shocking too. As one of the founders of the field, the Danish physicist Niels Bohr, said, "Anyone who is not shocked by quantum theory has not understood it."[1]

Does matter exist?

Quantum physics casts doubt on the most fundamental assumption of materialism – that is, on the very *existence* of matter. At least it suggests that matter doesn't exist in the way we normally understand it, as something solid and definite, which is always in a specific place and governed by laws of causality. At a microcosmic level, solid matter seems to dissolve away. Rather than being a hard, solid entity like a tiny billiard ball, a particle of matter is a whirl of energy manifesting itself as a particle at that particular moment. The contemporary physicist Paul Davies and the science writer John Gribbin have written that with the advent of quantum physics, "Newton's deterministic machine was replaced by a shadowy and paradoxical conjunction of waves and particles, governed by the laws of chance, rather than the rigid rules of causality… [S]olid matter dissolves away, to be replaced by weird excitations and vibrations of invisible field energy."[2] One of the originators of quantum theory, Max Planck, stated similarly: "As a man who has devoted his whole life to the most clear-headed science, to the study of matter, I can tell you as a result of my research about atoms this much: There is no matter as such."[3]

As Paul Davies and John Gribbin suggest, in the microcosmic realm there is sometimes no distinction between waves and particles – that is, entities may appear to be both a particle and a wave at the same time. There seems to be something more fundamental than both particles and waves, which can manifest itself as either. This particle/wave duality was demonstrated by one of the most famous (and most frequently repeated) experiments in quantum physics – the "double-slit" experiment, in which light passes through two slits in a plate and "lands" on another screen behind it. The light lands on the screen as particles, but, strangely, the particles also create an "interference pattern" that can only arise from them being waves. The pattern – a "fringe" of different stripes – arises where waves would overlap. (Variations of the experiment have found that each particle of light passes through only one slit, whereas a wave would pass through both of them. Even more strangely, it has also been found that if the experimenter detects which particular slit the particles pass through, then the wave-like interference pattern does not appear.) So, in the experiment light has properties of both waves and matter. Originally carried out in 1927, the experiment has been repeated many times in recent years with many other different types of particles, including much-larger entitles such as molecules made up of over 800 atoms.[4]

The role of the observer

From a common-sense materialist perspective, it seems that we are "in here" looking at a world "out there", and that we can only influence the external world through physical touch or force. In classical Newtonian physics, consciousness has no place. The world operates without it, like a machine. It would exist in exactly the same way if we weren't there to be conscious of it.

This explains why Newtonian physics is so closely affiliated to the materialist worldview outlook – it confirms our impression that the world exists independently of us, and that consciousness is just an incidental by-product of brain functioning.

But in quantum physics this gulf fades away. It is not possible to separate the observer from what he or she is observing. Our own consciousness participates in the world. For example, whether an entity manifests itself as a particle or a wave appears to depend on our expectations, and how an experiment is designed. If a quantum physicist sets up an experiment to observe particles, particles will be observed; if the experiment is designed to observe waves, then waves will appear. In a sense, nothing is a particle or a wave until it is observed, or until an experiment is conducted. In the quantum world, the observer always interacts, changing the behaviour of particles through their acts of observation and measurement. As already noted (in Chapter 8), Pascual Jordan, one of the founders of quantum physics, stated, "Observations not only disturb what has to be measured, they produce it."[5] Another relevant finding here – which also seems to lend support to the concept of precognition – is from the physicist John Wheeler's "delayed-choice" experiment, which showed that whether a photon behaves as a wave or a particle depends on a measurement decision made at a later time.[6]

The quantum concept of the "collapse of the wave function" suggests that a particle is never in a specific place until it is measured or observed. Until you actually look at it, a particle doesn't even have a definite existence. Like so much in quantum physics, this seems to defy rationality. In a sense, a particle is everywhere and nowhere, in a wide range of possible places. But the act of looking at a particle makes it real and locates it in a specific place. This "collapses" the wave function, bringing the particle out of possibility and into reality. Another aspect of this vagueness is what the early quantum physicist Werner Heisenberg called the "uncertainty principle", which shows that you can never know both the position and the momentum of a particle at the same time. The more specifically you try to measure its momentum, the less accurately you'll be able to measure its position.

The importance of the observer in co-creating what we perceive as "reality" was emphasized by John Wheeler, who wrote: "Nothing is more important about the quantum principles than this, that it destroys the concept of the world as 'sitting out there', with the observer safely separated from it... In some strange way the universe is a participatory universe."[7]

Entanglement and non-locality

The participatory nature of the world as revealed by quantum physics is illustrated even more clearly by the concepts of entanglement and non-locality. Once particles interact with one another, or when they emerge from the same source, they behave as if they are connected, spinning together and balancing each other's random fluctuations. No matter how far apart they are, they move in harmony.

This was an important finding in early quantum physics, and it has been repeatedly demonstrated by experiments ever since. For example, in 2015 researchers at the National Institute of Standards and Technology (NIST) in the United States created pairs of identical photons and sent them to two different particle detectors, nearly 660 feet (200 metres) apart. (The detectors

were around 430 feet [130 metres] from the original source
of the particles.) They found that the behaviour of the two
was highly correlated, and that the correlations appeared to be
unlimited and unrelated to any local factors. The researchers
calculated that the odds against the correlations being caused by
chance, or by local factors, was about one in 170 million.[8]

In 2017 a team of Chinese scientists demonstrated,
rather impressively, that entangled particles maintained their
link 870 miles (1,400 kilometres) away from each other.
Using a "quantum satellite" floating high above the Earth,
the researchers also found that it was possible to "teleport"
photons across this distance. Since the 1990s, physicists
have known that the link between entangled particles can
be used to transmit quantum information from one place to
another. The information associated with a photon can be
transmitted to a photon it is entangled with, so that the second
photon effectively takes on the identity of the first. Previous
experiments had "only" been able to teleport photons up
to a distance of 62 miles (100 kilometres), but the Chinese
researchers were able to massively extend this because for most
of their journey to the satellite the photons were travelling
through a vacuum.[9]

No one is sure how particles can apparently communicate
over large distances – this is just one of the things that, from the
standpoint of standard classical physics, makes no sense at all. As
the physicist Nick Herbert has put it, "A non-local connection
links up one location with another without crossing space,
without decay, and without delay."[10] According to Herbert,
the connections have three important characteristics: they are
unmediated (since no signal passes between them), unmitigated
(in the sense that the strength of the connection doesn't fade
over distance) and immediate (they happen instantaneously). In
this way, they seem to transcend both space and time.

One important thing to consider here is that during the Big
Bang – and shortly after it – all the particles in the universe were
in close proximity to each other, and must have interacted with
each other. As a result, all particles, and all of the larger entities

that are made up of particles, have always been twinned with each other. That means they are all, in theory, interconnected across distance and time, which suggests that the whole universe is entangled. *Everything* in the universe is non-local.

In other words, there is no separation between any particles – in the same way that there is no separation between the observer and the observed. As the contemporary physicist Carlo Rovelli puts it, "All variable aspects of an object exist only in relation to other objects. It is only in interactions that nature draws the world. In the world described by quantum mechanics there is no reality except in the relations between physical systems."[11] Or in the words of the British physicist David Bohm, "Ultimately, the entire universe… has to be understood as a single undivided whole, in which analysis into separate and independent existent parts has no fundamental status."[12]

The quantum effects in the macrocosmic world

Some of the findings we've looked at so far seem more like science fiction than scientific reality (I wouldn't be surprised if you thought of *Star Trek* while reading about "quantum teleportation"!). In quantum physics, terms like "paranormal" and "normal" cease to have any real meaning. It is difficult to see why some materialists believe that "paranormal" phenomena like telepathy and precognition are impossible, when the world is full of scientifically well-established but inexplicable phenomena which directly contravene the principles of Newtonian physics. If entanglement, "quantum tunnelling" and "quantum teleportation" are real, then why shouldn't telepathy be?

Some observers deal with the anomalies of quantum physics by pretending that they have no relevance to the macrocosmic world. After all, we live quite happily in the macrocosmic world of classical physics, which functions according to common-sense principles, and so is easy to make sense of. So why should we bother about what happens at a microcosmic level? That's a different level of reality, separate and independent from the one we live in.

This attitude is hypocritical in that the normal approach of science is to explain large-scale phenomena in terms of smaller ones. This is the standard reductionist approach. For example, materialism often attempts to explain human behaviour in terms of genes, or human experience (or human consciousness) in terms of the activity of brain cells. But for many materialists, the reductionist enterprise is halted at the level of molecules. There's no attempt to go further down, to the level of atoms and below – to neutrons, electrons and quarks.

Actually, it is easy to understand why reductionism is halted at this level: because the reductionist enterprise is used to support a materialist worldview, and at the level of the atom the principles of this worldview begin to fall apart. But it certainly seems disingenuous to make a microcosmic "cut-off point" and disregard everything below it.

The attitude is also illogical. There is no gulf between the classical macrocosmic world and the quantum microcosmic world. They are not two separate realms. The microcosmic world *constitutes and informs* the macrocosmic world, in the same way that the black dots with different shades make up a photograph. What is the case (or appears to be the case) for the quantum world is also the case for the macrocosmic world we live in. The quantum world *has* to be taken account of.

Moreover, as we already have seen briefly (in chapters 8 and 9), quantum biology has clearly shown that quantum effects occur in the macrocosmic world. As the Oxford physicist Vlatko Vedral wrote in 2011, "Until the past decade, experimentalists had not confirmed that quantum behaviour persists on a macrocosmic scale. Today, however, they routinely do. These effects are more pervasive than anyone ever suspected... We can't simply write [quantum effects] off as mere details that matter only on the very smallest scales."[13]

Quantum biology attempts to solve some very puzzling biological processes in terms of quantum effects such as entanglement and quantum tunnelling. We discussed adaptive mutation in these terms in Chapter 9, and here I will briefly summarize three other processes that can be explained in terms of quantum effects. The first is photosynthesis, which has always puzzled biologists because energy from the sun always unerringly finds the most direct route to the "reaction centre" of a cell of a leaf, through a mass of molecules. Experiments have shown that packets of energy from the sun don't do this randomly. They don't behave as if they are particles, trying to find a single route to the reaction centre. They behave like waves in quantum physics, spread out over space in superposition, trying out all possible routes at the same time and choosing the most direct one.[14]

Similarly, it is well known that in some enzymes protons move to different molecules through the process of quantum tunnelling. The protons pass through the solid boundaries of the molecule, a little like a ghost walking through a wall. They somehow teleport from one position to another, without physically moving across space. This massively speeds up chemical reactions (by a trillionfold, according to some estimates) and is essential to life.[15]

As a third and final example, there is a theory that quantum effects can help to explain the mystery of how some migrating birds navigate. For example, every winter thousands of robins fly 2,000 miles (3,220 kilometres) from Scandinavia to the Mediterranean, and it is known that they do this without reference to landmarks, ocean currents or the position of the sun or the stars. They navigate by detecting variations in the Earth's magnetic field. On the surface of it, this seems impossible because the Earth's magnetic field is incredibly weak – about 100 times weaker than a typical fridge magnet. But there is evidence that the robins have an "inner chemical compass" that works via the quantum process of entanglement. It is thought that there is a protein in the bird's eye that contains electrons which are entangled with particles in the Earth's magnetic field, which enables them to orientate themselves in relation to it.[16]

All of this makes it clear that it is impossible to separate the quantum world from the macrocosmic everyday world. Quantum weirdness doesn't just occur inside tiny particles; it is happening all around us. Quantum effects are everywhere, and quantum biology will no doubt uncover many more processes where they play an important role. We can no longer just derive our worldview from the principles of classical Newtonian physics and decide that quantum physics has no relevance to us.

This strategy of "separating off" the quantum world is probably the main reason why the findings of quantum physics haven't had much impact on our culture's materialist worldview, even though they have been known for many decades and are so well established. How could materialism have become so dominant when quantum physics has always shown that its basic assumptions are flawed? Materialists must always have known, if only unconsciously, that the findings of quantum physics threaten materialism and so they chose to ignore them (or perhaps to convince themselves that one day a grand theory of some kind would be developed that would suddenly make rational sense of quantum strangeness). Materialists have continued to look at the world as if quantum physics never happened, still holding to the strictly Newtonian common-sense view of reality.

But it is no longer possible to do this. We have to allow the findings of quantum physics to influence our worldview.

Quantum physics and panspiritism

From the perspective of panspiritism, there are two ways of looking at quantum physics. One is what you could call a cautious negative perspective, and the other is a more speculative positive one.

The cautious negative perspective is simply that quantum physics shows that materialism is inadequate as an explanation of the world. As we've seen, a number of the basic principles of materialism are contravened by quantum physics. These include the assumption that matter is the primary element of the universe; that consciousness is a by-product of matter; that the world exists "out there", separate and independent from our consciousness; and that the world consists of solid particles that are separate from one another and can only interact by physically touching or moving each other.

In the process of casting doubt on materialism, quantum physics has revealed the world to be a much more mysterious and complex place than we could ever imagine. It makes it clear that we cannot, in William James's phrase, close our "accounts with reality"[17] and assume that we have a good understanding of how the world works. The world as we perceive it appears to have very little in common with the world as it actually is. (This suggests another reason why there has been a tendency to ignore the implications of quantum physics: it threatens our sense of certainty, our sense that we have a reasonably good understanding of the world, and the feeling of control over the world that this certainty brings.)

This is the very least that quantum physics tell us. But if you're in a more speculative mood, it is it possible to go beyond this and suggest that quantum physics actually *supports* the existence of an all-pervading spiritual force, and even that it reveals certain aspects of it. Certainly, this is the conclusion that many quantum physicists have reached. As was mentioned earlier (in Chapter 2), many of the originators of quantum physics – such as Max Planck, Werner Heisenberg and Nils Bohr – were effectively mystics, who believed that consciousness (or spirit) is the primary reality of the universe, and the essence of all things, so that everything is interconnected. As Max Planck wrote: "I regard consciousness as fundamental. I consider matter as derivative of consciousness."[18] Or as David Bohm noted: "A rudimentary consciousness is present even at the level of particle physics."[19]

No doubt these physicists adopted a "panspiritist" perspective partly because their findings made them aware that materialist explanations of reality were inadequate, which encouraged them to investigate and adopt alternative views of reality. (For example, some of the original quantum physicists, such as Werner Heisenberg, were avid readers of mystical and spiritual texts.) But at the same time, there are clearly aspects of quantum physics that suggest such an outlook. The concepts of entanglement and non-locality suggest an underlying connectedness, which could be provided by the manifestation of an all-pervading fundamental consciousness. The important role of human consciousness in quantum physics points to this too, suggesting that in some way we share the same nature as everything we observe, and that we participate in and co-create the world we perceive. These seem to be hints of the universe's fundamental oneness and our fundamental oneness with the universe. At the same time, the fact that matter seems to emerge from something more fundamental than itself – and that both waves and particles are expressions of something more fundamental – suggests an underlying force of some kind. What could this force be? Is it too much of a stretch to suggest that it could be a fundamental force of consciousness, or a fundamental spirit-force?

I'm personally included to be cautious here. Since quantum physics tell us very little about the nature of this underlying reality, I think it would be a leap of faith to say that it is the same all-pervading spiritual force that has been referred to throughout this book. As we have seen, quantum physics is so counterintuitive and difficult to make sense of that it seems reckless to draw any firm conclusions.

However, many quantum physicists – who, after all, surely understood the field better than anyone else – certainly didn't hesitate to think in terms of an underlying and fundamental mind or consciousness. The British physicist James Jeans wrote that the universe "begins to look more like a great thought than a great machine" that exists "in the mind of some eternal spirit".[20] But perhaps the clearest expression of this worldview was given by Max Planck, who wrote:

> All matter originates and exists only by virtue of a force which brings the particle of an atom to vibration

and holds this most minute solar system of the atom together... We must assume behind this force the existence of a conscious and intelligent Spirit. This Spirit is the matrix of all matter.[21]

Again, the spiritual tone of these statements takes one by surprise, because we're so used to thinking in terms of a duality between spirituality and science. Perhaps the most significant aspect of quantum physics – and the reason why it is so important in relation to the argument of this book – is that quantum physics shows that you cannot separate science and spirituality. They cannot exist without each other. They are one and the same.

CHAPTER 12
THE SPIRITUAL UNIVERSE: MOVING BEYOND MATERIALISM

The late-16th and early 17th century was a dangerous time to be a free thinker, or a scientist. One man who paid the price for questioning his culture's metaphysical paradigm was the Italian Giordano Bruno. He was a true Renaissance man, an intellectual giant who was equally philosopher, poet, mathematician and cosmologist. (He was also a Dominican monk for 11 years.) Bruno accepted Copernicus's theory that the sun was the centre of the solar system, held a panspiritist view that all nature was alive with spirit and also believed in reincarnation. As far as the Church leaders were concerned, he contravened many of their core principles and so undermined their authority. In 1593 he was tried for heresy, charged with denying several core Catholic doctrines, found to be an impenitent and burned at the stake in 1600.

Galileo was another Italian free thinker who suffered at the hands of the Church. Galileo's astronomical investigations convinced him that the Earth was not the centre of the universe, and that our planet revolved around the sun. However, the Catholic Church saw "heliocentrism" as heretical. As a result, Galileo spent the latter part of his life under house arrest and his books were banned.

The main reason why the Church authorities were so hostile towards scientists and free thinkers was probably because they knew – if only unconsciously – that their metaphysical paradigm was under serious threat. Their brutal punishments were an attempt to hold back cultural change, like a corrupt leader who embarks on a murderous rampage as his grip on power is fading. But they were fighting a futile battle, of course. The shift was underway, and it was inevitable that their simplistic, biblical-based worldview would be superseded.

And I believe there are parallels with our present cultural situation.

In the last chapter of this book – in addition to highlighting some of the most important points we've touched on and summarizing my overall argument – I am going to suggest that at the moment a cultural shift is occurring and the metaphysical paradigm of materialism is fading away. It is important – for the future of our own species and for our planet as a whole – that this shift comes to full fruition, and that the materialist paradigm is transcended by a spiritual worldview.

Materialism under threat

Just like the Church in the 17th century, materialism is under threat. Its tenets and assumptions are no longer viable, and a new paradigm is emerging to replace it. And materialism is responding aggressively to this challenge, just as the Church did. Obviously, materialists are much less brutal than the Church, and don't go to the lengths of imprisoning or killing their enemies. But some materialists are reacting with a similar closed-mindedness and hostility.

There are three main ways in which metaphysical paradigms react to existential threats: by becoming more rigidly dogmatic; by punishing "heretics"; and by ignoring (or explaining away or suppressing) unwelcome evidence. This is still the way that fundamentalist religions maintain themselves in the midst of 21st-century secular culture. Fundamentalist Christians and Muslims – or any religious sect or cult, for that matter – have extremely rigid and specific beliefs and tenets, which every adherent has to wholly accept. Fundamentalists inculcate fear by ostracizing and punishing anyone who strays from these beliefs, and they attempt to restrict the availability of any evidence that contravenes the beliefs. And unfortunately, some adherents of the belief system of materialism respond in a similar way to challenges to their worldview. Free thinkers who question any of the tenets of materialism are accused of being pseudo-scientific. Particularly if free thinkers accept the existence of psi phenomena and even investigate them, they may find it difficult to get funding for research, to publish their work in journals, present it at conferences or get an academic post at a university. Free thinkers may be ridiculed, have their Internet pages doctored and have their videos taken down from the Internet (as happened to Rupert Sheldrake in 2013, when his TED talk was deleted at the behest of prominent American sceptics.) And as we saw in Chapter 8, some fundamentalist materialists may even purposely suppress or manipulate evidence that supports psi phenomena.

It's important to remember that there are powerful psychological factors at play here. Some materialists may not even realize that they are being closed-minded and prejudiced. Their behaviour is rooted in a powerful psychological need for certainty and control. As described in Chapter 1 (and touched on again in the last chapter), as a belief system materialism provides a coherent explanatory framework that makes sense of life. It seems to offer convincing answers to many of the "big questions" of human life, and as a result it provides people with a sense of orientation and certainty that alleviates doubt and confusion.

As we saw earlier, understanding is in a sense *over*standing. Feeling that we understand how the world works gives us a sense of authority. Rather than feeling subordinate to the mysterious and chaotic forces of nature, we feel that we overstand the world, in a position of power. As we saw in the last chapter in relation to quantum physics, to admit that phenomena exist which we can't fully understand or explain, and that the world is stranger than we can conceive, weakens our sense of power and control.

This issue of power applies to all belief systems, but in the case of materialism the feeling of control is enhanced by an attitude of dominance towards the rest of the natural world. Because we experience ourselves as separate from nature, and because we experience nature as fundamentally inanimate and mechanistic, we subconsciously feel entitled to dominate and exploit it. (As mentioned in Chapter 1 – and my discussion of the exploration of the psychological roots of materialism – there is very possibly a colonial element to this, which pictures nature as uncharted territory that it is our duty to explore and conquer, in much the same way as our European ancestors explored and conquered "new" continents.)

All of this means that once a person feels that their belief system is under threat they usually react with great hostility. To accept that the principles of your worldview are false – and that you have much less power and control over the world than you thought – is a dangerous step into the unknown.

This is the situation that materialism finds itself in now. It is being undermined, and in the process of being superseded. And some of its adherents are reacting exactly as both history and psychology would predict.

The failure of materialism

Materialism is a worldview extrapolated from some of the principles of science, so it is ironic that the main way in which materialism is being undermined is through some of the findings of modern science. As the last chapter showed, in a sense this has always been true of quantum physics and it is even more the case now that we are beginning to realize – through the field of quantum biology – how big a part quantum effects play in the macrocosmic world.

But in a more general sense, since the heyday of materialism in the second half of the 20th century the likelihood that it can serve as an adequate explanation of the world has dwindled. The confidence of genetic materialists was dented by the Human Genome Project – the mapping of the human genome. As previously noted in Chapter 4, human beings have far fewer genes than expected, and a large proportion of them are shared with other life forms. The role of genes was also found to be less significant than expected, suggesting that some inheritable traits are largely unrelated to genes. The project was originally predicted to bring about a revolution in healthcare by showing how common diseases were caused by the inheritance of faulty genes. As the director of the project, Francis Collins, stated at its onset, "In the next five to seven years, we should identify the genetic susceptibility factors for virtually all common diseases."[1] But it was found that faulty genes don't have a major role in predisposing us to disease. (As noted in Chapter 9, the project also helped to dispel the simplistic notion that there are "genes for" specific traits and human characteristics.)

In fact, the whole of this book has made it clear that confidence in the explanatory power of materialism is fading away. The confidence of neurological materialists has been dented by the lack of progress in attempts to explain consciousness in neurological terms. Chapter 3 showed that the more we look into the way the brain functions, the less likely it seems that there is any direct causal link between consciousness and brain activity. And as chapters 4 and 5 revealed, a diverse range of phenomena such as neuroplasticity and the placebo effect have cast a great deal of doubt on the simplistic assumption that the mind is nothing more a projection of the brain. As a result, it is highly unlikely that the Connectome Project – designed to map the human brain – will offer much as an explanation of human behaviour (although it may, of course, have other benefits).

In addition, the summaries in chapters 6–8 of recent research into near-death experiences, psychic phenomena and awakening experiences (or spiritual experiences) make it clear that there are many forms of human experience that cannot be accounted for in materialist terms, and that cannot be explained away as hallucinations or delusions. In Chapter 9 it was shown how recent concepts such as "adaptive mutation" have cast doubt on the Neo-Darwinian view of evolution, adding to the argument that evolution has not been a random process. As we saw in Chapter 10, recent research in anthropology and archaeology undermines the claims of evolutionary psychology and further undermines Neo-Darwinism by showing that it is invalid as an explanation of human behaviour.

At the same time as materialism is failing, post-materialist perspectives are beginning to flourish. Of course, these two developments aren't unconnected – the failure of materialism has made alternative perspectives seem more valid and has encouraged theorists to adopt them. For example, the failure to explain consciousness in neurological terms has led to a renewed interest in panpsychism and idealism, both of which suggest that consciousness is a fundamental quality of the universe. The journey of the neuroscientist Christof Koch (see Chapter 2) epitomizes this. Koch began as a materialist, convinced that consciousness could be explained in terms of its "neural correlates", but eventually he came to believe that this was a forlorn hope and adopted panpsychism instead.

In a similar way, there appears to be a growing consensus that the materialist approach to physical and mental health – one that treats the body as a machine and sees mental disorders as neurological problems that can be "fixed" through drugs – is seriously flawed. Increasing numbers of medical practitioners are moving towards more holistic approaches, with a greater awareness of the importance of environmental and psychological factors, and of how the mind can influence the health of the body. In particular, there is a growing awareness of the lack of efficacy of psychotropic drugs such as anti-depressants, and a movement towards more holistic therapies, such as cognitive behavioural therapy, mindfulness and ecotherapy (the use of contact with nature as a form of therapy to alleviate psychological problems).

And in a more general sense the increasing popularity of spiritual practices and paths suggests a cultural movement towards post-materialism. Spiritual development begins with a sense that there is "more to life" than the materialist worldview tells us, with an intuition that we – and all living beings – are more than simply biological machines whose consciousness is a kind of hallucination, and that natural phenomena are more than simply objects which we share the world with. Spirituality is an attempt to break the cultural trance of materialism and to transcend the limited, delusory vision that it is associated with.

The limits of awareness

I hope that this book has made it clear that the spiritual (or panspiritist) perspective offers a coherent and viable explanation of the world. Once we allow for the existence of an all-pervading spiritual force, a host of problematic phenomena – such as consciousness, the influence of the mind over the body, near-death experiences, psychic phenomena, even the weirdness of quantum physics, and so on – become more comprehensible. It is impossible to make any sense of these phenomena except in terms of a spiritual vision of the universe. Trying to explain them without the concept of a spiritual force is like trying to explain heat and light without the sun.

At the same time, it is important to remember that there is are limits to our awareness, and there are some things that even a panspiritist perspective may never be able to explain. There are many "big questions" that seem to be beyond our capability to answer. We will probably never be able to understand why photons of light can also be waves. We may never know whether the universe has an end or goes on forever. We may never understand dark energy or dark matter. We may also never be able to answer the "big question" of developmental biology: how does a single fertilized cell develop into a complex multi-cellular life form? (After their success in "breaking the genetic code" in the 1960s, some of the world's leading molecular biologists turned their attention to this problem, believing that it would only take them a decade or two to come up with a basic answer. And they're still looking now.) In many areas, it seems that the more we try to unravel nature's secrets, the more mystery we find. The deeper we look into the heart of things, the more riddles and puzzles we seem to uncover and the further we move away from any certainty.

Here it's important to remember the example I used in Chapter 8, of a sheep and its limited awareness. We are animals too, of course. Our awareness may be more intense than that of most other creatures, but that doesn't mean we are conscious to an absolute degree. This would be tantamount to suggesting that we are the end point of the whole evolutionary process, which is obviously ridiculous. Since our awareness is limited, there must be phenomena that are beyond our awareness and understanding.

This brings us back to the question of evolution. Eventually, assuming evolution continues, living beings with a more intense awareness than us will develop, and they will be aware of – and understand – more phenomena than us. They will have a more intense awareness of reality than us in the same way that we have a more intense awareness than a sheep – and their knowledge and understanding of the world will be correspondingly larger.

This is, incidentally, one way that spiritual paths and practices could help us. Because they have the effect of expanding our awareness, they could potentially enable us to become aware of phenomena that are presently hidden from us. This would be a different interpretation of the term "spiritual science", and a different approach to increasing our knowledge. The normal Western approach, in both science and philosophy, is to take the world we know as a "given" and examine and explore it in as much detail as possible. But the spiritual – and typically Eastern – approach is not to try to see in more detail but to try to *expand and intensify* our vision, so that we see with a greater range and subtlety. So, in theory, this is an approach we could follow to try to increase our knowledge and understanding, although it is unlikely that our knowledge would ever become absolute.

The tenets of panspiritism

If materialism only offers a limited and distorted vision of reality, what does panspiritism tell us about the world – and the universe – we live in, and about our own nature as human beings?

Close to the beginning of the book "The Ten Tenets of Materialism" were listed, and in a similar way I'm going to end the book by listing a number of aspects of the panspiritist worldview. This spiritual worldview is a metaphysical paradigm in the same way that materialism is, but the important point is that it is a *much more valid* paradigm, supported by much more evidence and with the capacity to include (and explain) a vast range of phenomena which don't fit into the materialist model. This inclusivity is one of the most important aspects of panspiritism. The materialist model can only function by *excluding* psychic phenomena, near-death experiences, spiritual experiences, the self-healing through mental influence, pure altruism and many other phenomena. This is the only way that such a narrow and rigid worldview can preserve itself – to ignore everything it can't explain. It's similar to how a small totalitarian state might try to preserve itself by barring any travel or media contact with other countries and pretending that the rest of the world doesn't really exist. But the panspiritist worldview has no need to exclude anything, which is a sure sign of its validity.

I'll follow roughly the same pattern as my list in Chapter 1, including elements of some of the original materialist tenets, to illustrate how they are contravened by panspiritism.

- Life did not come into being through the accidental interactions of certain chemicals, but did so as the result of the innate tendency of the universe – propelled by consciousness itself – to move towards greater complexity. At a certain point, when material entities (in the form of the first simple cells) reached a certain degree of complexity, they became able to receive and transmit universal consciousness. This was when life began.

- Evolution is not an accidental process. Once life forms had come into existence, evolution was impelled by the innate tendency of consciousness to generate greater complexity. More complex life forms enabled universal consciousness to express itself more intensely, manifesting itself as individual consciousness.

- Rather than being just biological machines, human beings are, both mentally and physically, expressions of spirit, or consciousness. You can say that our physical bodies are an external expression of universal consciousness, while our minds (or beings) are an inner expression. What we normally think of as "spirit" is universal consciousness as it is expresses itself inside us. Therefore – although in slightly different senses – both body and mind are "spiritual".

- As the previous point suggests, our personal consciousness – our subjective or inner life – is not generated by the brain. It is a fundamental universal quality, which our brains "receive" and canalize into our individual being.

- Mental phenomena cannot be reduced to neurological activity. There may be correlations between mental and neurological activity, but there is certainly not a one-way causal link. Neurological changes can affect mental experience, but mental experience can also bring about neurological changes.

- In a similar way, the mind exerts a powerful influence on the body. It can bring about healing and illness, and even change the structure of the body. This is because, while both mind and body are expressions of consciousness, the mind is a more subtle and intense manifestation of consciousness.

- Because consciousness is not produced by the brain, and because we are more than just physical stuff, we should be open to the possibility of some form of life after death. Consciousness will not come to an end when our brains and bodies die. In fact, evidence (for example, from near-death experiences and after-death communications) suggests an afterlife in which we continue to have a sense of individual identity.

- Human beings are not isolated entities, moving through the world in separation to one another. We share the same essence, and are therefore deeply interconnected. We express (and become aware of) this connection through empathy, compassion and altruism.
- The world does not exist "out there" in separation to us. Our own consciousness is deeply interconnected with it. We share the same essence as all things, and are therefore one with all things.
- Our normal state of awareness is limited and delusory, and does not provide us with an accurate perception of the world "as it is". In higher states of consciousness – or awakening experiences – we gain a more expansive and intense awareness, and attain a fuller and truer perception of reality.
- Paranormal phenomena such as telepathy and precognition do not contravene the laws of science. From the perspective of post-materialism, they are not only possible but natural. For example, since we share the same essential consciousness with other human beings, it is not surprising that we sometimes sense each other's thoughts and intentions (as in telepathy).

As these statements show, the spiritual worldview is so different to the materialist one that it's hard to believe that they are different interpretations of the *same* world. You could compare it to two paintings of the same landscape by two artists who have such different techniques and received such different impressions that you would never guess they shared the same subject.

The materialist worldview is bleak and barren; it tells us that life is fundamentally purposeless and meaningless, that we're just here for a few decades and it doesn't really matter what we do. It's no wonder that so many people respond to this by just trying to have as much of a "good time" as they can, to take all they can get from the world without worrying about the consequences, or else by diverting themselves with distractions like television or numbing themselves with alcohol and other drugs. It seems inevitable that people should try to take refuge from the bleakness of materialism by living materialistically, treating themselves to as much fun and as many consumer products as they can afford, or trying to build up their wealth and status and power.

However, the spiritual worldview tells us that the universe is not a bleak vacuum. It tells us that the nature of the universe is bliss. This is because the fundamental nature of consciousness itself is bliss. We have seen evidence for this many times throughout this book – for example, in near-death experiences and high-intensity awakening experiences, when individual consciousness becomes more intense and subtle, seeming to merge powerfully with universal consciousness to bring a profound sense of peace and euphoria. (This is expressed succinctly in the Hindu teaching that the nature of *brahman* is *satchitananda* – or "being–consciousness–bliss".) This bliss is inside us too, because we are individual expressions of consciousness. As countless spiritual teachers have told us, there is no need to search for happiness outside us – in material goods or pleasures and power – because well-being is our true nature.

The spiritual worldview also tells us that human nature is not essentially malevolent but benign. Selfishness and cruelty are not natural, they are aberrational. They only occur when we lose our sense of connection; when our fundamental oneness is obscured by an aberrational sense of ego-separateness. In essence, we exist in co-operation rather than competition, and we are altruistic rather than selfish. In essence we are one. We are, literally, each other.

Finally, the spiritual worldview tells us that our lives are meaningful and purposeful. As stated earlier, the purpose of our lives is the same as that of evolution itself – to deepen our sense of connection to others through empathy and altruism, to uncover as much of our innate potential as we can, and to expand and intensify our awareness. The purpose of our lives is, you might say, *self*-evolution.

A transformation of consciousness

At the present time, the issue of self-evolution is highly significant. It is imperative that we undergo as much self-evolution as possible – not simply for our own sake, but for the sake of the whole human race.

Because the metaphysical paradigm of materialism has had – and continues to have – so many disastrous effects, it is essential for our culture as a whole to adopt a post-materialist spiritual worldview as soon as possible. Ultimately – as many Native American Indian leaders pointed out to the Europeans who came to ransack their continent – materialism leads to environmental destruction. As an approach to life, it is hopelessly out of harmony with nature. It encourages the reckless plundering of the Earth's resources, the hopeless and ceaseless search for satisfaction through consumer goods and hedonistic adventures, and even the exploitation and oppression of other human beings. As such, materialism is unsustainable. Unless it is superseded, it is likely that we will experience a catastrophic cultural breakdown and major ecological devastation – potentially even the extinction of our species.

Moving beyond materialism means daring to question the received wisdom of our culture and examining the assumptions we have absorbed from it. It means being brave enough to risk ridicule and ostracism from fundamentalists who are fighting a futile battle to maintain an outmoded worldview. But perhaps more than anything else, moving beyond materialism means experiencing the world in a different way.

In Chapter 1 I emphasized that, at the most fundamental level, materialism stems from our perception of the world. It stems from the perception of the world as an inanimate place, and of natural phenomena as inert objects. It stems from our experience of ourselves as entities who live inside our own mental space in separation from the world and other human beings and living beings. In order to transcend materialism, it is therefore essential that we transcend this mode of perception. Moving beyond materialism means becoming able to perceive the vividness and sacredness of the world around us. It means transcending our sense of separateness so that we can experience our connectedness with nature and other living beings.

I mentioned earlier that spiritual practices and paths can help us by expanding our awareness, thereby increasing our potential knowledge of the world. But they can actually provide us with an even bigger benefit by helping us to transcend the limited awareness that gives rise to the materialist worldview. This is the primary purpose of spiritual practices and paths: to "undo" the psychological structures that create our automatic vision of the world and our sense of separation. Spirituality wakes us up, opens us to the aliveness and sacredness of nature, and reconnects us to the world. When we experience the world in this way, we truly move beyond materialism.

This is the most important issue of our time. We don't need to explore the outer world in any more detail; we need to turn inside and explore our own being. New technologies to further manipulate the world aren't so important now; it's more urgent for us to make use of "spiritual technologies" to help us to expand our awareness and as a result attain a new vision of the world.

Because every human being is interconnected, the more we evolve as individuals, the more we will help our whole species to evolve. As we individually transcend the "sleep" vision that has given rise to materialism, we will be helping our whole species to do the same. And eventually this limited vision will fade away, like a mirage, and we will collectively remember who we really are, and where we really are. We will no longer perceive ourselves as soulless biological machines, but as radiant and purposeful manifestations of spirit. We will no longer perceive the world as a soulless physical machine, but as a radiant and meaningful manifestation of spirit. We will sense our oneness with the world, and treat it with the care and respect it deserves.

In addition to explaining the world, spirituality may actually help to save it.

ACKNOWLEDGEMENTS

Many thanks to Penny Sartori, Paul Marshall, Michael Duggan, Paul Kieniewicz and Joseph Bray for their comments on individual chapters of this book. Thanks also to the authors of the seminal text *Irreducible Mind*, which has been an inspiration to me since I first read it several years ago. (Sometimes, when friends or colleagues asked me what my new book was about, I replied, "I'm trying to write a popular version of *Irreducible Mind*." In some chapters – particularly Chapter 5 – I also made direct use of some of the research cited in the book.) Thanks to my friends David Lawton and Paul Marshall (again) for providing philosophical advice and support over various lunches. Finally, I would like to express my gratitude to Tjasa Seljak, who helped prepare the notes and bibliography.

NOTES

Introduction

1. I actually prefer the term *trans*-materialism because it emphasizes that this perspective moves beyond materialism rather than simply coming after it. But the term post-materialism is already current. In fact, a "post-materialist science" movement has recently been initiated by a large group of scientists, including the neuroscientist Mario Beauregard and the psychologist Lisa Miller.

Chapter 1
The Origins of Materialism: When Science Turns into a Belief System

1. Newton, 2018.
2. Huxley, 1874, p. 577.
3. Maudsley, 1879, p. 667.
4. Fromm, 1974, p. 311.
5. Bacon, 2018.
6. Nietzsche, 1967, p. 23.
7. Dawkins, 1998, p. 6.
8. *ibid*.
9. Compton and Hoffman, 2012.
10. Frankl, 2006.
11. See my article "Two Ways of Seeing the World" (Taylor, 2016a).
12. In Crowley, 1994, p. 35.

Chapter 2
The Spiritual Alternative

1. In Skrbina, 2017, p. 26.
2. In Otto, 1960, p. 60.

3. Bruno, 1998, p. 44.
4. Chalmers, 1995, p. 83.
5. Means and Wolf, 1995.
6. In Griffiths, 1976, p. 21.
7. Munro, 1962.
8. Levy-Bruhl, 1965, p. 17.
9. Ingold, 2000, p. 67.
10. Mascaro, 1990, p. 75.
11. In Brockelman, 1999, p. 27.
12. In McLuhan, 1971, p. 36.
13. *ibid.*, p. 61.
14. In Smith, 2009, pp. 158–160.
15. Mascaro, 1990, p. 117.
16. Gottlieb, 2004, p. 39.
17. Versulius, 1994, p. 34.
18. In McLuhan, 1971, p. 45.
19. Suzuki, 2000, p. 34.
20. Griffiths, 1976, p. 25.
21. Wordsworth, 1994, p. 648.
22. Whitman, 1980, p. 220.
23. Lawrence, 1994, p. 587.
24. Schroedinger, 1983, pp. 31–32.

Chapter 3
The Riddle of Consciousness

1. Crick, 1994, p. 3.
2. Crick and Koch, 1990, p. 263.
3. Reiss and Marino, 2001.
4. Ramachandran and Hirstein, 1997.
5. Tononi, 2018, p. 2.
6. Another possibility put forward by the physicists Roger Penrose and Stuart Hameroff is that consciousness is related to quantum processes in the brain, involving microtubules (tiny protein structures inside nerve cells). According to this view, the microtubules serve as "quantum computers", enabling a collapse of the wave function, causing quantum coherence and bringing consciousness into being. However, critics have argued that this just seems to be explaining one mystery in terms of another – that is, the mystery of consciousness in terms of the mystery of quantum physics.
7. McGinn, 1993, p. 60.
8. Koch, 2014, p 28.
9. Forman, 1998, p. 185.

Chapter 4
The Primacy of Mind: Puzzles of the Mind and Brain

1. Robinson, 1995, p. 306.
2. Mural et al., 2002.
3. Hall, 2010.
4. Tononi, 2018, p. 4.
5. Kastrup, 2013.
6. In Weir, 2012, p. 30.
7. Pandya et al., 2012, p. 640.
8. Healy, 2015.
9. *ibid.*
10. Kirsch and Sapirstein, 1998.
11. Puras, 2017.
12. Burton et al., 2002.
13. Feuillet et al., 2007.
14. Thomson, 2014.
15. Smith and Sugar, 1975.
16. Kelly et al., 2007.

Chapter 5
How the Mind Can Change the Brain and the Body: More Puzzles of the Mind and Brain

1. Huxley, 1874, p. 292.
2. Maguire et al., 2000.
3. Draganski et al., 2006.
4. Hölzel et al., 2011.
5. Mason et al., 2017.
6. Levine et al., 1978.
7. Kelly et al., 2007.
8. Kallmes et al., 2009.
9. Sihvonen et al., 2013.
10. Wartolowska et al., 2014
11. Esdaile, 1846, p. xxiv.
12. Abdeshahi et al., 2013.
13. Levine et al., 1978; Kelly et al., 2007.
14. Kelly et al., 2007.
15. *ibid.*
16. *ibid.*
17. *ibid.*
18. Reeves et al., 2007.
19. Edwards et al. 2010.
20. Kelly at al., 2007.

21. Fried et al., 1951.
22. Schopbach et al., 1952.
23. Phillips and Smith, 1990.

Chapter 6
The Puzzle of Near-Death Experiences

1. In Fontana, 2005, p. 387.
2. In Fenwick, 1995, p. 95.
3. In Lorimer, 1990, p. 86.
4. Parnia et al., 2014.
5. Van Lommel et al., 2001.
6. Greyson, 2003.
7. Knoblauch et al., 2001.
8. Kelly et al., 2007.
9. Taylor, 2011, p. 170.
10. Grey, 1985, p. 97.
11. *ibid.*
12. In Keim, 2013.
13. Kelly et al., 2007, p. 416.
14. Sacks, 2012.
15. Sartori, 2014, p. 112.
16. Rivas et al., 2016, p. 11.
17. *ibid.*, p. 21.
18. *ibid.*, p. 22.
19. Sartori, 2014.
20. Sabom, 1982.
21. Kelly et al., 2007.
22. Sartori, 2014.
23. Irwin and Watt, 2007, p. 172.
24. Sartori, 2014, p. 179.
25. Van Lommel, 2010, p. xvii.
26. A triple-blind study is a randomized experiment in which the treatment or intervention is unknown to the research participant, the researcher and the individual who assesses the outcomes of the experiment.
27. Barušs and Mossbridge, 2016, p. 22.
28. Rock et al., 2014.
29. Carter, 2012.
30. Sartori, 2014, pp. 85–86.
31. Osis and Haraldsson, 2012.
32. Greeley, 1987; Lindstrom, 1995.
33. Stevenson, 1993.
34. Kelly et al., 2007, p. 598.

Chapter 7
Waking Up: The Puzzle of Awakening Experiences

1. Stange and Taylor, 2008.
2. Taylor and Egeto-Szabo, 2017, p. 61.
3. *ibid.*, p. 54.
4. Taylor, 2011, p. 10.
5. Taylor, 2010, p. 10.
6. Taylor, 2012.
7. Taylor and Egeto-Szabo, 2017, p. 54.
8. *ibid.*
9. Forman, 1998.
10. Prabhavananda and Manchester, 1957, p. 51.
11. Taylor and Egeto-Szabo, 2017, p. 56.
12. *ibid.*, p. 49.
13. Maslow, 1994.
14. Persinger, 1983; Ramachandran and Blakesee, 1998.
15. Newberg and D'Aquili, 2000.
16. Kelly et al., 2007.
17. Aean-Stockdale, 2012; Devinksy and Lai, 2008.
18. Aaen-Stockdale, 2012.
19. Karnath et al., 2001; Kelly et al., 2007.
20. Mascaro, 1990, p. 53.
21. *ibid.*, p. 61.
22. In Happold, 1986, p. 279.

Chapter 8
Keeping the Account Open:
The Puzzle of Psychic Phenomena

1. American Psychological Association, 2013.
2. Lange et al., 2000.
3. Sheldrake, 2003.
4. Taylor, 2007.
5. Judd and Gawronski, 2011, p. 406.
6. In Carr, 2011, p. 2.
7. *ibid.*
8. *ibid.*
9. Bem et al., 2014.
10. Rhine, 1997.
11. Honorton and Ferrari, 1989.
12. Mossbridge et al., 2012.
13. Radin, 2006.
14. Dalton, 1997.
15. Bem and Honorton, 1994.
16. Morris et al., 1993.

17. Sheldrake and Smart, 2000.
18. Radin, 2006.
19. Baptista and Derakhshani, 2014.
20. Ritchie et al., 2012.
21. Prinz et al., 2011.
22. Begley and Ellis, 2012.
23. Whitehead, 1948, p. 129.
24. Hofstadter, 2011.
25. Penrose, 1999, p. 574.
26. Davies, 1977, p. 221.
27. In Hoffmann, 1972, p. 258.
28. Sheehan, 2006, p. vii.
29. Jordan, 1957, p. 16.
30. Cramer, 2016.
31. In Jammer, 1974, p. 151.
32. Brookes, 2017.
33. De Beauregard, 1975, p. 101.
34. In Shieber, 2004, p. 88.
35. Ehrenwald, 1978, p. 138.
36. Penman, 2008.
37. Utts, 1996, p. 3.
38. Hyman, 1996, p. 43.
39. Delgado-Romero and Howard, 2005, p. 298.
40. Sheldrake, 1999; Sheldrake and Smart, 2000.
41. Wiseman et al., 1998; Sheldrake and Smart, 2000.
42. Wilber, 1982.

Chapter 9
Complexity and Consciousness: Puzzles of Evolution

1. Cepelewicz, 2017.
2. Wills and Carter, 2017.
3. Carter and Wills, 2018.
4. Guseva et al., 2017.
5. Fernández-García et al., 2017.
6. Nagel, 2012. p. 6.
7. *ibid.*
8. *ibid.*
9. *ibid.*, p. 16.
10. In Zammito, 2013, p. 84.
11. *ibid.*, p. 86.
12. Capra, 1996.
13. Grassé, 1977, p. 97.
14. *ibid.*, p. 96.
15. See Kauffman, 1995; Reid, 2007.
16. Conway Morris, 2006, p. 327.

17. Foster, 2000.
18. Al-Khalili and McFadden, 2014.
19. Darwin, 1859, p. 62.
20. Kropotkin, 2018.
21. Margulis and Sagan, 1997, p. 29.
22. Thornhill, 2001.
23. For a more detailed discussion see my article "How Valid is Evolutionary Psychology?" (Taylor, 2014).
24. Andrews and Thomson, 2009.
25. Humphrey, 1983; 2011.
26. Humphrey and Metzinger, 2017.

Chapter 10
Why Do Selfish Genes Behave So Unselfishly?
The Puzzle of Altruism

1. Samuels, 2017.
2. Wilkinson, 2017.
3. Buckley, 2017.
4. Dawkins, 1976, p. 66.
5. Boyer, 2001, p. 299.
6. Taylor, 2018.
7. Haas and Piscitelli, 2013.
8. Hassan, 1980.
9. Haas and Piscitelli, 2013.
10. Ferguson, 2013.
11. Of course, some contemporary tribal groups are warlike, but these groups generally don't follow the same, simple, "immediate return" way of life as our prehistoric ancestors. Some historical tribal groups – such as the Plains Indians – also became much more aggressive due to the transgressions of European colonists. In more recent times, there are many Amazonian tribes – such as the Jivaro and Yanomamo – who are extremely aggressive, which may also be due to centuries of attack and intrusion by colonists. Another possibility, supported by archaeological evidence, is that Amazonian tribes are the descendants of a civilization that disappeared during the 17th century. See my book *The Fall* for further details.
12. Fry and Söderberg, 2014.
13. Dyble et al., 2015.
14. Diamond, 1987, p. 64.
15. Dawkins, 1998, p. 212.
16. This connection has been identified by the American psychologist Daniel Batson, who has developed an "empathy-altruism" hypothesis to explain acts of pure altruism.
17. Compton and Hoffman, 2012.

18. Haidt, 2002, p. 864.
19. Schopenhauer, 1966, p. 379.
20. In Lowney, 2006, p. 258.

Chapter 11
Quantum Questions: Mysteries of the Microcosm

1. In Wheatley, 1999, p. 23.
2. Davies and Gribbin, 2007, p. 14.
3. Planck, 1944.
4. Eibenberger et al., 2013.
5. In Jammer, 1974, p. 151.
6. Wheeler, 1983.
7. In Mehra, 1973, p. 244.
8. Shalm et al., 2015.
9. Liao et al., 2017; Ren et al., 2017.
10. Herbert, 1987, p. 214.
11. Rovelli, 2015, p. 115.
12. Bohm, 1983, p. 174.
13. Vedral, 2011, p. 39.
14. Al-Khalili and McFadden, 2014.
15. *ibid.*
16. *ibid.*
17. James, 1928, p. 388.
18. Planck, 1931.
19. In Skrbina, 2017, p. 262.
20. Jeans 1937, p. 137.
21. Planck, 1944.

Chapter 12
The Spiritual Universe: Moving Beyond Materialism

1. Collins, 2000.

BIBLIOGRAPHY

Aaen-Stockdale, C, "Neuroscience for the Soul" in *The Psychologist*, 25(7),
    pp. 520–523, 2012

Abdeshahi, SK et al., "Effect of Hypnosis on Induction of Local Anaesthesia,
    Pain Perception, Control of Haemorrhage and Anxiety During Extraction
    of Third Molars: A Case-control Study" in the *Journal of Cranio-Maxillo-
    Facial Surgery*, 41, pp. 310–315, 2013

Al-Khalili, JS and McFadden, JJ, *Life on the Edge: The Coming of Age of
    Quantum Biology*, Bantam Press, London, 2014

American Psychiatric Association, *Diagnostic and Statistical Manual of Mental
    Disorders*, (5th ed.), American Psychiatric Publishing, Washington, DC, 2013

Andrews, PW and Thomson Jr, JA, "The Bright Side of Being Blue: Depression
    as an Adaptation for Analyzing Complex Problems" in *Psychological
    Review*, 116(3), pp. 620–654, 2009

Bacon, F, *Novum Organum*, 2018 (Retrieved 6/2/18 from http://www.
    constitution.org/bacon/nov_org.htm)

Ball, Philip, "Physics of Life: The Dawn of Quantum Biology" in *Nature*, 474,
    pp. 272–274, 2011

Baptista, J and Derakhshani, M, "Beyond the Coin Toss: Examining Wiseman's
    Criticisms of Parapsychology" in *Journal of Parapsychology*, 78(1),
    pp. 56–79, 2014

Baruš, I and Mossbridge, J, *Transcendent Mind: Rethinking the Science of
    Consciousness*, American Psychological Association, Washington, DC, 2016

Begley, CG and Ellis, LM, "Drug Development: Raise Standards for Preclinical
    Cancer Research" in *Nature*, 483, pp. 531–533, 2012

Bem, D and Honorton, C, "Does Psi Exist? Replicable Evidence for an Anomalous
    Process of Information Transfer" in *Psychological Bulletin*, 115, pp. 4–18, 1994

Bem, DJ, "Feeling the Future: Experimental Evidence for Anomalous
    Retroactive Influences on Cognition and Affect" in *Journal of Personality
    and Social Psychology*, 100, pp. 407–425, 2011

Bem, D, Tressoldi, PE, Rabeyron, T and Duggan, M, "Feeling the Future:
    A Meta-Analysis of 90 Experiments on the Anomalous Anticipation
    of Random Future Events", 2014 (Available at SSRN: http://ssrn.
    com/abstract=2423692 or http://dx.doi.org/10.2139/ssrn.2423692
    Bibcode:2011Natur.474..272B. doi:10.1038/474272a)

Blackmore, SJ, *In Search of the Light: The Adventures of a Parapsychologist*, Prometheus, Amherst, NY, 1996

Bohm, DJ, *Wholeness and the Implicate Order*, Ark Paperbacks, London, 1983

Boyer, P, *Religion Explained*, Vintage, London, 2001

Brockelman, P, *Cosmology and Creation: The Spiritual Significance of Contemporary Cosmology*, Oxford University Press, New York, 1999

Brookes, JC, "Quantum Effects in Biology: Golden Rule in Enzymes, Olfaction, Photosynthesis and Magnetodetection" in *Proceedings of the Royal Society*, 473, 31 May 2017 (Available at https://doi:10.1098/rspa.2016.0822)

Bruno, G, *Cause, Principle and Unity: And Essays on Magic*, Cambridge University Press, Cambridge, 1998

Buckley, C, "Man Is Rescued by Stranger on Subway Tracks" in *New York Times*, 2017 (Retrieved 17/7/17 from http://www.nytimes.com/2007/01/03/nyregion/03life.html)

Burton, H, et al., "Adaptive Changes in Early and Late Blind: A fMRI study of Braille Reading" in *Journal of Neurophysiology*, 87, pp. 589–607, 2002

Buss, DM and Shackelford, TK, "Human Aggression in Evolutionary Psychological Perspective" in *Clinical Psychology Review*, 17(6), pp. 605–619, 1997

Capra, F, *The Web of Life*, Anchor Books, New York, 1996

Carr, B, "Heresies and Paradigm Shifts" in *The Network Review*, Spring 2011, pp. 2–4, 2011

Carter, C, "Persistent Denial: A Century of Denying the Evidence" in Krippner, S and Friedman, H (eds), *Debating Psychic Experience*, Praeger, Santa Barbara, CA, pp. 77–110, 2010

Carter, C, *Science and the Afterlife Experience: Evidence for the Immortality of Consciousness*, Inner Traditions, Rochester, New York, 2012

Carter, CW and Wills, PR, "Interdependence, Reflexivity, Fidelity, Impedance Matching, and the Evolution of Genetic Coding" in *Molecular Biology and Evolution*, 35(2), pp. 269–286, 2018 (Available at https://doi.org/10.1093/molbev/msx265)

Cepelewicz, J, "Life's First Molecule Was Protein, Not RNA, New Model Suggests" in *Quanta Magazine*, 12 November 2017 (Available at https://d2r55xnwy6nx47.cloudfront.net/uploads/2017/11/lifes-first-molecule-was-protein-not-rna-new-model-suggests-20171102.pdf)

Chalmers, DJ, "The Puzzle of Conscious Experience" in *Scientific American*, 273(6), pp. 80–86, 1995

Collins, F, Transcript of Interview with Collins, F., Director of the Human Genome Project at the National Institutes of Health, *Religion Ethics Newsweekly*, 2000 (Available at http://www.pbs.org/wnet/religionandethics/2000/06/16/transcript-bob-abernethys-interview-with-dr-francis-collins-director-of-the-human-genome-project-at-the-national-institutes-of-health/15204/)

Compton, WC and Hoffman, E, *Positive Psychology: The Science of Happiness and Flourishing*, (2nd ed.), Wadsworth, Belmont, CA, 2012

Conway Morris, S, *Life's Solution: Inevitable Humans in a Lonely Universe*, Cambridge University Press, Cambridge, 2006

Cordovero, M, *The Palm Tree of Deborah*, trans. L Jacobs, Sepher-Hermon Press, New York, 1976

Cramer, JG, *The Quantum Handshake: Entanglement, Nonlocality and Transactions*, Springer, Heidelberg, NY, 2016

Crick, F and Koch, C, "Towards a Neurobiological Theory of Consciousness" in *Seminars in the Neurosciences*, 2, pp. 263–275, 1990

Crick, F, *The Astonishing Hypothesis*, Simon & Schuster, New York, 1994

Crowley, V, *Phoenix to a Flame: Pagan Spirituality in the Western World*, Thorsons, London, 1994

Dalton, K , "Exploring the Links: Creativity and Psi in the Ganzfeld" in *Proceedings of Presented Papers*, Parapsychological Association, 40th Annual Convention, pp. 119–134, 1997

Davies, P and Gribbin, J, *The Matter Myth: Dramatic Discoveries That Challenge our Understanding of Physical Reality*, Simon & Schuster, New York, 2007

Darwin, C, *On the Origin of Species by Natural Selection, Or the Preservation of Favored Races in the Struggle for Life*, Murray, London, 1859

Davies, P, *Space and Time in the Modern Universe*, Cambridge University Press, Cambridge, 1977

Dawkins, R, *The Selfish Gene*, Oxford University Press, Oxford, 1976

Dawkins, R, *Unweaving the Rainbow: Science, Delusion and the Appetite for Wonder*, Houghton Mifflin Harcourt, Boston, 1998

De Beauregard, CO, "Quantum Paradoxes and Aristotle's Twofold Information Concept" in Oteri, L (ed.), *Quantum Physics and Parapsychology*, Parapsychology Foundation, New York, pp. 91–102, 1975

De Waal, F, *Primates and Philosophers: How Morality Evolved*, Princeton University Press, Princeton, NJ, 2009

Delgado-Romero, EA and Howard, GS, "Finding and Correcting Flawed Research Literatures" in *The Humanistic Psychologist*, 33, pp. 293–303, 2005

Dennett, DC, *Consciousness Explained*, Little, Brown and Co., Boston, 1991

Devinsky, O and Lai, G, "Spirituality and Religion in Epilepsy" in *Epilepsy Behaviour,* 12(4), pp. 636–643, 2008 (Available at https://doi:10.1016/j.yebeh.2007.11.011. Epub 2 January 2008)

Diamond, J, "The Worst Mistake in the History of the Human Race" in *Discover*, pp. 64–66, 1987

Draganski, B, et. al., "Temporal and spatial dynamics of brain structure changes during extensive learning" in *The Journal of Neuroscience*, 26(23), pp. 6314–6317, 2006

Dyble, M, et al., "Sex Equality Can Explain the Unique Social Structure of Hunter-Gatherer Bands" in *Science*, 15 May 2015, pp. 796–798, 2015

Edwards, TM, et al., "The Treatment of Patients With Medically Unexplained Symptoms in Primary Care: A Review of the Literature" in *Mental Health in Family Medicine*, 7(4), pp. 209–221, 2010

Ehrenwald, J, "Einstein Skeptical of Psi? Postscript to a Correspondence" in *Journal of Parapsychology*, 42, pp. 137–142, 1978

Eibenberger, S, et al., "Matter-wave Interference With Particles Selected From a Molecular Library With Masses Exceeding 10 000 Amu" in *Physical Chemistry Chemical Physics*, 15, pp. 14696–14700, 2013

Esdaile, J, *Mesmerism in India and its Practical Applications in Surgery and Medicine*, Longmans, Green and Longmans, London, 1846

Facco, E, et al., "Effects of Hypnotic Focused Analgesia on Dental Pain Threshold" in *International Journal of Clinical and Experimental Hypnosis*, 59(4), pp. 454–68, 2011

Fenwick, P and Fenwick, E, *The Truth in the Light*, Hodder Headline, London, 1995

Ferguson, RB, "The Prehistory of War and Peace in Europe and the Near East" in Fry, DP (ed.), *War, Peace, and Human Nature: The Convergence of Evolutionary and Cultural Views*, Oxford University Press, Oxford, pp. 191–240, 2013

Fernández-García, C, Coggins, A and Powner, M, "A Chemist's Perspective on the Role of Phosphorus at the Origins of Life" in *Life*, 7(3), p. 31, 2017

Feuillet, L, Dufour, H and Pelletier, J, "Brain of a White-collar Worker" in *The Lancet*, 370, p. 262, 2007 (Available at https://DOI:10.1016/S0140-6736(07)61127-1)

Fontana, D, *Is There an Afterlife?*, O Books, Winchester, 2005

Forman, Robert, "What Does Mysticism Have to Teach Us About Consciousness?" in *Journal of Consciousness Studies*, 5(2), pp. 185–201, 1998

Foster, PL, "Adaptive Mutation: Implications for Evolution" in *Bioessays*, 22, pp. 1067–1074, 2000 (Available at https://doi:10.1002/1521-1878(200012)22:12<1067::AID-BIES4>3.0.CO;2-Q)

Frankl, VE, *Man's Search for Meaning*, Beacon Press, Boston, 2006

Fried, PH, Rakoff, AE, Schopbach, RR and Kaplan, AJ, "Pseudocyesis: A Psychosomatic Study in Gynaecology" in *Journal of American Medical Association*, 14(2), pp. 129–134, 1951

Fromm, E, *The Anatomy of Human Destructiveness*, Jonathan Cape, London, 1974

Fry, DP and Söderberg, P, "Myths About Hunter-Gatherers Redux: Nomadic Forager War and Peace" in *Journal of Aggression, Conflict and Peace Research*, 6(4), pp. 255–266, 2014

Gottlieb, RS, *This Sacred Earth: Religion, Nature, Environment*, (2nd ed.), Routledge, London, 2004

Grassé, PP, *Evolution of Living Organisms*, Academic Press, New York, 1977

Greeley, AM, "Hallucinations Among the Widowed" in *Sociology and Social Research*, 71(4), pp. 258–265, 1987

Grey, M, *Return From Death*, Arkana, London, 1985

Greyson, B, "Incidence and Correlates of Near-Death Experiences in a Cardiac Care Unit" in *General Hospital Psychiatry*, 25(4), pp. 269–276, 2003

Griffiths, B, *Return to the Centre*, Collins, London, 1976

Guseva, E, Zuckermann, R and Dill, K, "Foldamer hypothesis for the growth and sequence differentiation of prebiotic polymers" in *Proceedings of the National Academy of Sciences*, 114(36), pp. 7460–7468, 2017 (Available at https://doi.org/10.1073/pnas.1620179114)

Haas, J and Piscitelli, M, "The Prehistory of Warfare: Misled by Ethnography" in Fry, DP (ed.), *War, Peace, and Human Nature*, pp. 168–190, Oxford University Press, New York, 2013

Haidt, J, "The Moral Emotions" in Davidson, RJ, Scherer, K and Goldsmith, HH (eds), *Handbook of Affective Sciences*, Oxford University Press, Oxford, pp. 852–870, 2002

Hall, SS, "Revolution Postponed" in *Scientific American*, 303, pp. 60–67, 2010

Happold, FC, *Mysticism: A Study and Anthology*, Pelican, London, 1986

Hassan, FA, "Prehistoric Settlements Along the Main Nile" in Williams, MAJ and Faure, H (eds), *The Sahara and the Nile: Quaternary Environments and Prehistoric Occupation in Northern Africa*, pp. 421–450, Balkema, Rotterdam, 1980

Healy, D, "Serotonin and Depression: The Marketing of a Myth" in *British Medical Journal*, 350, 2015 (Available at https://doi.org/10.1136/bmj.h1771)

Heisenberg, W, *The Physicists' Conception of Nature*, Hutchinson, London, 1958

Heisenberg, W, *Physics and Beyond*, Allen & Unwin, London, 1971

Herbert, N, *Quantum Reality: Beyond the New Physics*, Anchor Books, Garden City, NY, 1987

Hoffmann, B, *Albert Einstein: Creator and Rebel*, New American Library, New York, 1972

Hofstadter, D, "A Cutoff for Craziness" in *New York Times*, 2011 (Available at http://www.nytimes.com/roomfordebate/2011/ 01/06/the-esp-study-when-science-goespsychic/a-cutoff-for-craziness. Accessed 25/1/18)

Honorton, C and Ferrari, DC, "'Future Telling': A Meta-Analysis of Forced-Choice Precognition Experiments, 1935–1987" in *Journal of Parapsychology*, 53, pp. 281–308, 1989

Hobbes, T, *Leviathan*, Continuum International Publishing Group, London, 2006

Hölzel, BK, Carmody, J, Vangel, M, Congleton, C, Yerramsetti, SM, Gard, T and Lazar, SW, "Mindfulness Practice Leads to Increases in Regional Brain Gray Matter Density" in *Psychiatry Research: Neuroimaging*, 191(1), pp.36–43, 2011 (Available at http://www.sciencedirect.com/science/article/pii/S092549271000288X-cr0005)

Humphrey, N and Metzinger, T, "A Self Worth Having", 2017 (Retrieved 3/1/17 from https://www.edge.org/conversation/nicholas_humphrey-a-self-worth-having)

Humphrey, N, *Consciousness Regained: Chapters in the Development of Mind*, Oxford University Press, Oxford, 1983

Humphrey, N, *Soul Dust: The Magic of Consciousness*, Princeton University Press, Princeton, NJ, 2011

Huxley, TH, "On the Hypothesis That Animals are Automata and its History" in *Fortnightly Review*, n.s. 16, pp. 555–580, 1874

Hyman, R, "Evaluation of a Program on Anomalous Mental Phenomena" in *Journal of Scientific Exploration*, 10(1), pp. 31–58, 1996

Ingold, T, *The Perception of the Environment*, Routledge, London, 2000

Irwin, HJ and Watt, CA, *An Introduction to Parapsychology*, (5th ed.), McFarland & Company Inc., London, 2007

James, W, *The Varieties of Religious Experience: A Study of Human Nature*, Longmans, Green and Co., New York, 1902/1928

Jammer, M, *The Philosophy of Quantum Mechanics*, Wiley, New York, 1974

Jeans, J, *The Mysterious Universe*, Macmillan, New York, 1937

Jeans, J, *Physics and Philosophy*, Cambridge University Press, Cambridge, 1942/2009

Jefferson, W, *The World of Chief Seattle: How Can One Sell the Air?*, Native Voices Books, Summertown, TN, 2001

Jordan, P, "New Trends in Physics" in *Proceedings of Four Conferences of Parapsychology Studies*, Parapsychology Foundation, New York, 1957

Judd, CM and Gawronski, B, "Editorial Comment" in *Journal of Personality and Social Psychology*, 100(3), p. 406, 2011 (Available at https:// doi:10.1037/0022789)

Kallmes, DF, et al., "A Randomized Trial of Vertebroplasty for Osteoporotic Spinal Fractures" in *The New England Journal of Medicine*, 361(6): pp. 569–579, 2009

Kaptchuk, TJ, et al., "Placebos Without Deception: A Randomized Controlled Trial in Irritable Bowel Syndrome" in *PLoS One*, 5(12), p. 1559, 2010 (Available at https://doi:10.1371/journal.pone.0015591)

Karnath, H, Ferber, S and Himmelbach, M, "Spatial Awareness is a Function of the Temporal Not the Posterior Parietal Lobe" in *Nature*, 411, pp. 950–953, 21 June 2001 (Available at https://doi:10.1038/35082075)

Kastrup, B, *Why Materialism is Baloney: How True Skeptics Know There is no Death and Fathom Answers to Life, the Universe, and Everything*, Iff Books, Winchester, 2014

Kauffman, S, *At Home in the Universe: The Search for Laws of Self-Organization and Complexity*, Oxford University Press, Oxford, 1995

Keim, B, "Consciousness After Death: Strange Tales From the Frontiers of Resuscitation Medicine" in *Wired*, 2013 (Retrieved 13/8/2017 from https://www.wired.com/2013/04/consciousness-after-death/)

Kelly, EF, et al., *Irreducible Mind: Toward a Psychology for the 21st Century*, Rowman & Littlefield, Lanham, MD, 2007

Kirsch, I and Sapirstein, G, "Listening to Prozac But Hearing Placebo: A Meta-Analysis of Antidepressant Medication" in *Prevention and Treatment*, 1(2), 1998

Koch, C, "Is Consciousness Universal?" in *Scientific American Mind*, 25, pp. 26–29, 2014

Knoblauch, H, Schmied, I and Schnettler, B, "Different Kinds of Near-Death Experience: A Report on a Survey of Near-Death Experiences in Germany" in *Journal of Near-Death Studies*, 20, pp. 15–29, 2001

Kropotkin, PA, *Mutual Aid: A Factor of Evolution*, CreateSpace Independent Publishing, 2018 (Retrieved 18/11/17 from http://www.public-library.uk/ pdfs/7/883.pdf)

Lange, R, Irwin, HJ and Houran, J, "Top-Down Purification of Tobacyk's Revised Paranormal Belief Scale" in *Personality and Individual Differences*, 29, pp. 131–156, 2000

Lawrence, DH, *Complete Poems*, Penguin, London, 1994

Levine, JD, Gordon, NC and Fields, HL, "The Mechanisms of Placebo Analgesia" in *The Lancet*, 312(8091), pp. 654–657, 1978 (Available at http://www.sciencedirect.com/science/article/pii/ S0140673678927629)

Levy-Bruhl, L, *The Soul of the Primitive*, Allen & Unwin, London, 1965

Liao, S, et al., "Satellite-to-Ground Quantum Key Distribution" in *Nature*, 549, pp. 43–47, 7 September 2017 (Available at https://doi:10.1038/nature23655)

Lindstrom, TC, "Experiencing the Presence of the Dead: Discrepancies in 'The Sensing Experience' and Their Psychological Concomitants" in *Omega*, 31, pp. 11–21, 1995

Lorimer. D, *Whole in One*, Arkana, London, 1990

Lowney, C, *A Vanished World: Muslims, Christians, and Jews in Medieval Spain*, Oxford University Press, New York, 2006

Lucretius, *The Nature of Things*, trans. AE Stallings, Penguin, London, 2007

Maguire, EA, et al., "Navigation-Related Structural Change in the Hippocampi of Taxi Drivers" in *Proceedings of National Academy of Science*, 97(8), pp. 4398–4403, 2000

Mascaro, J (ed. and trans.), *The Upanishads*, Penguin, London, 1990

Maslow, A, *Religions, Values and Peak Experiences*, Arkana, New York, 1994

Mason, L, Peters, E, Williams, SC and Kumari, V, "Brain Connectivity Changes Occurring Following Cognitive Behavioural Therapy for Psychosis Predict Long-term Recovery" in *Translational Psychiatry*, 7(1), January 2017 (Available at https://doi:10.1038/tp.2016.263 https://www.nature.com/articles/tp2016263)

Margulis, L and Sagan, D, *Microcosmos: Four Billion Years of Microbial Evolution*, University of California Press, London, 1997

Maudsley, H, "Materialism and its Lessons" in *Popular Science Monthly*, 15, pp. 667–683, 1879

Mayer, EL, *Extraordinary Knowing: Science, Skepticism and the Inexplicable Powers of the Human Mind*, Bantam Dell, New York, 2007

McGinn, C, "Consciousness and Cosmology: Hyperdualism Ventilated" in Davies, M and Humphreys, GW (eds), *Consciousness*, Basil Blackwell, Oxford, pp. 155–177, 1993

McLuhan, TC, *Touch the Earth: A Self-Portrait of Indian Existence*, Abacus, London, 1971

Means, R and Wolf, MJ, *Where White Men Fear to Tread: The Autobiography of Russell Means*, St Martin's Press, New York, 1995

Mehra, J (ed.), *The Physicist's Conception of Nature*, Reidel, Dordrecht-Holland, 1973

Morris, R, Cunningham, S, McAlpine, S and Taylor, R, "Toward Replication and Extensions of Autoganzfeld Results" in *Proceedings of Presented Papers*, pp. 177–191, Parapsychological Association, 36th Annual Convention, Toronto, 1993

Mossbridge, J, Tressoldi, P and Utts, J, "Predictive Physiological Anticipation Preceding Seemingly Unpredictable Stimuli: A Meta-Analysis" in *Frontiers of Psychology*, 3, p. 390, 2012

Munro, NG, *Ainu Creed and Cult*, Columbia University Press, New York, 1962

Mural, RJ, et al., "A Comparison of Whole-genome Shotgun-derived Mouse Chromosome 16 and the Human Genome" in *Science*, 296(5573), pp. 1661–1671, 2002 (Available at DOI: 10.1126/science.1069193)

Nagel, T, *Mind and Cosmos: Why the Materialist Neo-Darwinian Conception of Nature is Almost Certainly False*, Oxford University Press, Oxford, 2012 (Available at https://doi.org/10.1093/acprof:oso/9780199919758.001.0001)

Newberg, A and D'Aquili, E, "The Neuropsychology of Religious and Spiritual Experience" in *Journal of Consciousness Studies*, 7(11–12), pp. 251–266, 2000

Newton, I, *The General Scholium to Isaac Newton's* Principia Mathematica, 2018 (Retrieved 2/1/2018 from https://isaac-newton.org/general-scholium/)

Nietzsche, F, *The Birth of Tragedy and the Case of Wagner*, trans. W Kaufmann, Vintage Books, New York, 1967

Nietzsche, F, *Thus Spoke Zarathustra*, trans. G Parkes, Oxford World's Classics, Oxford, 2005

Osis, K and Haraldsson, E, *At the Hour of Death*, (3rd ed.), Hastings House, Norwalk, CT, 2012

Otto, R, *The Idea of the Holy*, Penguin, London, 1960

Pandya, M, Altinay, M, Malone, DA and Anand, A, "Where in the Brain Is Depression?" in *Current Psychiatry Reports*, 14(6), pp. 634–642, 11 October 2012 (Available at http://doi.org/10.1007/s11920-012-0322-7)

Parnia, S, et al., "AWARE—AWAreness During REsuscitation—A Prospective Study" in *Resuscitation*, 85(12), pp. 1799–805, December 2014

Penman, D, "Could There Be Proof to the Theory That We're ALL Psychic?" in MailOnline, 28 January 2008 (Retrieved 13/9/17 from http://www.dailymail.co.uk/news/article-510762/Could-proof-theory-ALL-psychic.html)

Penrose, R, *The Emperor's New Mind: Concerning Computers, Minds, and the Laws of Physics*, Oxford University Press, Oxford, 1999

Persinger, MA, "Religious and Mystical Experiences as Artefacts of Temporal Lobe Function: A General Hypothesis" in *Perceptual and Motor Skills*, 57(3 Pt 2), pp. 1255–1262, 1983 (PMID 6664802. Available at https://doi:10.2466/pms.1983.57.3f.1255)

Phillips, DP and Smith, DG, "Postponement of Death Until Symbolically Meaningful Occasions" in *Jama*, 263, pp. 1947–1951, 1990

Planck, M, interview with *The Observer*, London, 25 January 1931

Planck, M, speech in Florence, Italy, 1944 (from Archiv zur Geschichte der Max-Planck-Gesellschaft, Abt. Va, Rep. 11 Planck, Nr. 1797)

Prabhavananda, S and Manchester, F (trans.), *The Upanishads: Breath of the Eternal*, Mentor, New York, 1957

Prinz, F, Schlange, T and Asadullah, K, "Believe It or Not: How Much Can We Rely on Published Data on Potential Drug Targets?" in *Nature Reviews: Drug Discovery*, 10(712), 31 August 2011

Puras, D, "Depression: Let's Talk About How We Address Mental Health", World Health Day, 7 April 2017 (Available at http://www.ohchr.org/EN/NewsEvents/Pages/DisplayNews.aspx?NewsID=21480&LangID=E)

Radin, D, *Entangled Minds*, Paraview/Pocket, New York, 2006

Ramachandran VS and Blakesee, S, *Phantoms in the Brain*, Fourth Estate, London, 1998

Ramachandran, VS and Hirstein, W, "Three Laws of Qualia: What Neurology Tells Us About the Biological Functions of Consciousness" in *Journal of Consciousness Studies*, 4(5–6), pp. 429–457, 1997

Reeves, R, Ladner, ME, Hart, RH and Burke, RS, "Nocebo Effects With Antidepressant Clinical Drug Trial Placebos" in *General Hospital Psychiatry*, 29(3), pp. 275–277, 2007

Reid, R, *Biological Emergences Evolution by Natural Experiment*, MIT Press, Cambridge, MA, 2007

Reiss D and Marino, L, "Mirror Self-Recognition in the Bottlenose Dolphin: A Case of Cognitive Convergence" in *PNAS*, 98(1), pp. 5937–5942, 2001

Ren, JG, "Ground-to-Satellite Quantum Teleportation" in *Nature*, 549, pp. 70–73, 7 September 2017 (Available at https://doi:10.1038/nature23675)

Rhine, JB, *Extra-Sensory Perception*, Branden, Boston, MA, 1934/1997

Ritchie, SJ, Wiseman, R and French, CC, "Failing the Future: Three Unsuccessful Attempts to Replicate Bem's 'Retroactive Facilitation of Recall' Effect" in *PLoS One*, 7, 2012 (Available at http://journals.plos.org.plosone/article?id=10.371/journal.pone.0033423)

Rivas, T, et al., *The Self Does Not Die: Verified Paranormal Phenomena from Near Death Experiences*, (1st ed.), IANDS Publications, Durham, NC, 2016

Robinson, DN, *An Intellectual History of Psychology*, (3rd ed.), University of Wisconsin Press, Wisconsin, 1995

Rock, A, et al., "Discarnate Readings by Claimant Mediums: Assessing Phenomenology and Accuracy under Beyond Doubt-Blind Conditions" in *Journal of Parapsychology*, 782, pp. 183–194, 2014

Rovelli, C, *Seven Brief Lessons on Physics*, Penguin Random House, London, 2015

Sabom, MB, *Recollections of Death*, Harper & Row, New York, 1982

Sacks, OW, "Seeing God in the Third Millennium" in *The Atlantic,* 12 December 2012 (Retrieved 17/8/17 from https://www.theatlantic.com/health/archive/2012/12/seeing-god-in-the-third-millennium/266134/)

Sahlins, M, "Notes on the Original Affluent Society" in Lee, RB and DeVore, I (eds), *Man the Hunter*, Aldine, Chicago, pp. 85–89, 1968

Samuels, G, "Manchester Terror Attack: Homeless Man 'Pulled Nails Out of Child's Face'" in *The Independent*, 23 May 2017 (Retrieved 13/7/17 from http://www.independent.co.uk/news/uk/home-news/manchester-terror-attack-homeless-man-pulled-nails-face-children-injured-arena-bombing-a7751656.html)

Sartori, P, *The Wisdom of Near-Death Experiences*, Watkins, London, 2014

Schopbach, RR, Fried, PH and Rakoff, AE, "Pseudocyesis, A Psychosomatic Disorder" in *Psychosomatic Medicine*, 14(129), 1952

Schopenhauer, A, *The World as Will and Representation*, Dover Publications, New York, 1966

Schroedinger, E, *My View of the World*, Ox Bow Press, Woodbridge, CT, 1983

Shalm, L, et al. , "Strong Loophole-free Test of Local Realism" in *Physical Review Letters*, 115(250402), 10 November 2015 (Available at https://arXiv:1511.03189)

Sheehan, DP (ed.), "Frontiers of Time: Retrocausation—Experiment and Theory" in *AIP Conference Proceedings, San Diego, California*, American Institute of Physics, Melville, New York, 2006 (Available at https://philpapers.org/rec/SHEFOT)

Sheldrake, R, "Commentary on a Paper by Wiseman, Smith, and Milton on the 'Psychic Pet' Phenomenon" in *Journal of the Society for Psychical Research*, 63(857), pp. 233–255, 1999

Sheldrake, R and Smart, P, "A Dog That Seems to Know When his Owner is Coming Home: Videotaped Experiments and Observations" in *Journal of Scientific Exploration*, 14(2), pp. 233–255, 2000

Sheldrake, R, *The Sense of Being Stared At: And Other Unexplained Powers of the Human Mind*, Crown Publishers, New York, 2003

Shieber, S, *The Turing Test: Verbal Behavior as the Hallmark of Intelligence*, MIT Press, Cambridge, MA, 2004

Sihvonen, R, et al., "Arthroscopic Partial Meniscectomy Versus Sham Surgery for Degenerative Meniscal Tear" in *New England Journal of Medicine*, 369, pp. 2515–2524, 2013

Skrbina, D, *Panpsychism in the West*, MIT Press, Cambridge, MA, 2017

Smith, A and Sugar, O, "Development of Above Normal Language and Intelligence 21 Years After Left Hemispherectomy" in *American Academy of Neurology*, 25(9), p. 813, 1975

Smith, H, *Tales of Wonder*, HarperOne, San Francisco, 2009

Stange, K and Taylor, S, "Relationship of Personal Cognitive Schemas to the Labeling of a Profound Emotional Experience as Religious-Mystical or Aesthetic" in *Empirical Studies of the Arts*, 26, pp. 37–49, 2008

Stevenson, I, "Birthmarks and Birth Defects Corresponding to Wounds on Deceased Persons" in *Journal of Scientific Exploration*, 7(4), pp. 403–410, 1993

Stuss, DT, Picton, TW and Alexander, MP, "Consciousness, Self-awareness, and the Frontal Lobes" in Salloway, SP, Malloy, PF and Duffy, JD (eds), *The Frontal Lobes and Neuropsychiatric Illness*, American Psychiatric Publishing, Inc., Arlington, VA, pp. 101–109, 2001

Suzuki, DT, *The Awakening of Zen*, Shambhala, Boston, MA, 2000

Taylor, S, *Making Time: Why Time Seems to Pass at Different Speeds and How to Control it*, Icon Books Ltd, Cambridge, 2007

Taylor, S, *Waking from Sleep: Why Awakening Experiences Occur and How to Make Them Permanent*, Hay House, London, 2010

Taylor, S, *Out of the Darkness: From Turmoil to Transformation*, Hay House, London, 2011

Taylor, S, "Spontaneous Awakening Experiences: Exploring the Phenomenon Beyond Religion and Spirituality" in *Journal of Transpersonal Psychology*, 44(1), pp. 73–91, 2012

Taylor, S, "How Valid is Evolutionary Psychology?" on the blog Out of the Darkness, 2014 (Available at https://www.psychologytoday.com/blog/out-the-darkness/201412/how-valid-is-evolutionary-psychology [Accessed 7 Dec. 2017])

Taylor, S, "Two Ways of Seeing the World" on the blog Out of the Darkness, 2016 (Available at https://www.psychologytoday.com/blog/out-the-darkness/201608/two-ways-seeing-the-world [Accessed 7 Nov 2017])

Taylor, S, "From Philosophy to Phenomenology: The Argument for a 'Soft' Perennialism" in *International Journal of Transpersonal Studies*, 35(2), pp. 17–41, 2016

Taylor, S, *The Leap: The Psychology of Spiritual Awakening*, New World Library, Novato, CA, 2017

Taylor, S, *The Fall: The Insanity of the Ego in Human History and the Dawning of a New Era*, (2nd ed.), O-Books, Ropley, Hampshire, 2018

Taylor, S and Egeto-Szabo, K, "Exploring Awakening Experiences: A Study of Awakening Experiences in Terms of their Triggers, Characteristics, Duration and Aftereffects" in *Journal of Transpersonal Psychology*, 49(1), pp. 45–65, 2017

Thomson, H, "Woman of 24 Found to Have No Cerebellum in Her Brain" in *New Scientist*, 10 September 2014 (Available at www.newscientist.com/ article/mg22329861-900-woman-of-24-found-to-have-no-cerebellum-in-her-brain)

Thornhill, R and Palmer, CT, *A Natural History of Rape: Biological Bases of Sexual Coercion*, MIT Press, Cambridge, MA, 2001

Tononi, G, "Consciousness and the Brain: Theoretical Aspects" in Adelman, G, and Smith, B (eds), *Encyclopedia of Neuroscience*, (3rd ed.), Elsevier, 2004 (Retrieved 3/6/2017 from https://jsmf.org/meetings/2003/nov/ consciousness_encyclopedia_2003.pdf)

Utts, JM, "An Assessment of the Evidence for Psychic Functioning" in *Journal of Scientific Exploration*, 10(1), pp. 3–30, 1996

Van Lommel, P, van Wees, R, Meyers, V and Elfferich, I, "Near-Death Experience in Survivors of Cardiac Arrest: A Prospective Study in the Netherlands" in *The Lancet*, 358, pp. 2039–2045, 2001

Van Lommel, P, *Consciousness Beyond Life: The Science of the Near-Death Experience*, HarperOne, New York, 2010

Vedral, V, "Living in a Quantum World" in *Scientific American*, 304(6), pp. 38–43, 2011

Versulius, A, *Sacred Earth: The Spiritual Landscape of Native America*, Inner Traditions, Rochester, NY, 1992

Wartolowska, K, et al., "Use of Placebo Controls in the Evaluation of Surgery: Systematic Review" in *British Medical Journal*, 348, pp. 32–53, 21 May 2014 (Available at https://www.ncbi.nlm.nih.gov/pmc/articles/PMC4029190/)

Weir, K, "The Roots of Mental Illness: How Much of Mental Illness can the Biology of the Brain Explain?" in *APA Monitor*, 43(6), p. 30, 2012

Wheatley, MJ, *Leadership and the New Science: Discovering Order in a Chaotic World*, Berrett-Koehler Publications, San Francisco, 1999

Wheeler, JA, "Law without Law" in JA Wheeler and WH Zurek (eds), *Quantum Theory and Measurement*, Princeton University Press, Princeton, NJ, pp. 182–213, 1983.

Whitehead, AN, *Essays in Science and Philosophy*, Philosophical Library, New York, 1948

Whitman. W, *Leaves of Grass*, Signet Books, New York, 1980

Wilber, K, (1982). "The Pre/Trans Fallacy" in *Journal of Humanistic Psychology*, 22(2), pp. 5–43, 1982

Wilkinson, D, "The NHS Heroes Who Came to Manchester's Aid in the City's Darkest Hour" in *Manchester Evening News*, 25 May 2017 (Retrieved 17/7/2017 from https://www.manchestereveningnews.co.uk/news/greater-manchester-news/nhs-heroes-who-came-manchesters-13094631)

Wills, PR and Carter, CW, "Insuperable Problems of the Genetic Code Initially Emerging in an RNA World" in *Biosystems*, 164, pp. 155–166, February 2018 (Available at https://164. 10.1016/j.biosystems.2017.09.006)

Wiseman, R, Smith, M and Milton, J, "Can Animals Detect When Their Owners are Returning Home? An Experimental Test of the 'Psychic Pet' Phenomenon" in *British Parapsychology*, 42, pp. 137–142, 1998

Wordsworth, W, *The Works of William Wordsworth*, Wordsworth Editions, Ware, Hertfordshire, 1994

Wrangham, RW and Peterson, D, *Demonic Males: Apes and the Origins of Human Violence*, Houghton Mifflin Harcourt, New York, 1996

Zammito, J, "The Nagel Flap: Mind and Cosmos" in *The Hedgehog Review*, 15(3), pp. 84–94, 2013

INDEX